Hugh Macmillan

Holidays on High Lands

Hugh Macmillan

Holidays on High Lands

ISBN/EAN: 9783337290108

Printed in Europe, USA, Canada, Australia, Japan

Cover: Foto ©Andreas Hilbeck / pixelio.de

More available books at **www.hansebooks.com**

HOLIDAYS ON HIGH LANDS;

OR,

Rambles and Incidents

IN SEARCH OF

ALPINE PLANTS.

BY THE

REV. HUGH MACMILLAN,

AUTHOR OF "BIBLE TEACHINGS IN NATURE," ETC.

London:
MACMILLAN AND CO.
1869.

PREFACE.

The following chapters may be regarded as popular studies in geographical botany. Although each is separate and distinct, they have all a common basis and bond of unity. Their aim is to impart a general idea of the origin, character, and distribution of those rare and beautiful Alpine plants which occur on the British hills, and which are found almost everywhere in Europe, Asia, Africa, and America, wherever there are mountain chains sufficiently lofty to furnish them with a suitable climate. In the first three chapters, the peculiar vegetation of the Highland mountains is fully described; while in the remaining chapters this vegetation is traced to its northern cradle in the mountains of Norway, and to its southern European termination in the Alps of Switzerland. All the excursions mentioned were made during intervals of relaxation from professional work extending over several summers. Instead of conveying the information I have to give regarding the plants gathered on these occasions, in technical language, in a formal treatise, I thought it better that it should appear in a setting of personal adventure, and be associated with descriptions of the natural scenery and the peculiarities of the human life in the midst of which the plants were found. By this method of treatment the subject will perhaps be made more interesting to a larger circle of readers.

H. M.

GLASGOW,
June 1, 1869.

CONTENTS.

CHAPTER I.
	PAGE
THE PLANTS ON THE SUMMITS OF THE HIGHLAND MOUNTAINS	1

CHAPTER II.
THE INTERMEDIATE OR HEATHER REGION 76

CHAPTER III.
A GARDEN WALL IN A HIGHLAND GLEN 112

CHAPTER IV.
A RAMBLE THROUGH NORWAY, THE CRADLE OF THE HIGHLAND FLORA 144

CHAPTER V.
THE SKJEGGEDAL-FOSS IN NORWAY 226

CHAPTER VI.
THE PASS AND HOSPICE OF THE GREAT ST. BERNARD . . 256

HOLIDAYS ON HIGH LANDS.

CHAPTER I.

THE PLANTS ON THE SUMMITS OF THE HIGHLAND MOUNTAINS.

A THOUGHTFUL man, standing beneath the silent magnificence of the midnight heavens, is more deeply impressed by what is suggested than by what is revealed. He cannot gaze upon the solitary splendour of Sirius, or the clustered glories of Orion, without a vague unuttered wish to know whether these orbs are inhabited, and what are the nature and conditions of existence there. A similar feeling of curiosity seizes us when we behold afar off the summits of a lofty range of mountains, lying along the golden west like the shores of another and a brighter world. Elevated far above the busy commonplace haunts of men, rearing their mystic heads into the clouds, they seem to claim affinity with the heavens, and, like the stars, to dwell apart, retiring into a more awful and sacred solitude than exists on the surface of the earth. We long for the wings of the eagle, to surmount in a moment all intervening obstacles, and reach the shores of that upper world, that we may know what strange arrangements of matter, what new forms of life, occur in a region so near to and so favoured of the skies. To many individuals destitute of the strength of limb and soundness of lung necessary to climb the mountain side, or chained hopelessly to the monotonous employments by which the daily bread is earned, this must ever be an unattainable enjoyment—in sight, and yet unknown. Even of the thousands of tourists who as duly as the autumn comes round swarm over the familiar Highland routes, very few turn aside to behold this great sight. Only a solitary adventurous pedestrian, smitten with the love of science, now and then cares to diverge from the beaten paths, from the region of coaches and extortionate hotel-keepers, to explore the primeval solitudes of the higher hills. For these and other reasons, a brief description of the characteristic vegetation of the Highland mountains may prove interesting and instructive to many. The information I have to lay before the reader has been acquired with much toil during many summer wanderings; but if it should be the means of opening up to any one the way to a new field of research and a new set of sensations, it will be the source of much satisfaction to me.

Mountains exercise a peculiar and powerful fascination over the imagination. They transport us out of the fictitious atmosphere of civilization, and the cramping air of the world of taskwork, into the region of poetry and freedom. Among their serene and quiet retreats, the fevered, conventional life, brought face to face with the purity and the calm of nature, reverts to its primitive simplicity, the mind recovers its original elasticity, and the heart glows with its native warmth. Every individual finds in them something to admire, and to suit the tendencies of his mind. To the patriot, they are the monuments of history, which have attracted to themselves, by kindred sympathy, some of the most remarkable events that have diversified the life of nations—guardians of liberty, whose high, embattled ridges form an impenetrable rampart against the invading foe, and nourish within their fastnesses a hardy race, free as their own wild winds. To the poet, they are the altars of nature, on which the golden-robed sun offers his morning and evening sacrifice—footstools of God, before which his soul kneels, hushed in awe and reverence. To the philosopher, they are the theatres in which the mightiest forces of nature are seen in intensest action,—the storehouses in which are treasured up all the sources of earth's beauty and fertility. While to the devotional mind they are types of the stability of the Christian promises,—emblems of the Infinite, the Eternal, and the Unchangeable.

The fascination which mountains exercise extends to all that is connected with them. Their own sublimity and grandeur are reflected, as it were, upon all their productions; and the lowliest object that hides under their shadow, or is nourished by their soil, acquires from that circumstance an importance which does not intrinsically belong to it. Hence the peculiar charm which all botanists find in the pursuit of Alpine botany. The plants which grow upon the rugged sides, and the bleak storm-scalped summits of the mountains, cannot generally be compared, in point of variety and beauty of colouring, and luxuriance of growth, with the flowers of the plains. They are, for the most part, tiny plants, that, among their leaves of light, have no need of flowers—harmonizing in all their characters with their dreary habitats, and claiming apparently a closer affinity to the grey lichens and the brown mosses among which they nestle, than to their bright sisters of the valleys. But by their comparative rarity, by the magnificent and almost boundless prospects obtained from their elevated haunts, and by the exhilarating nature of the mountain breezes and scenery, they are surrounded by a halo of interest far exceeding that connected with woodland flowers; and a glowing enthusiasm is felt in their collection which cannot be experienced in the tamer and less adventurous pursuit of lowland botany.

The Highland mountains occupy but a very subsidiary position among the great mountain ranges of the earth. The highest peak in which they culminate does not reach the line of perpetual snow; no avalanche thunders over their precipices to bury the villages at their base in ruins; no glacier brings eternal winter down from his elevated throne into the midst of green corn-fields and cultivated valleys, or yawns in dangerous crevasses across the traveller's path; and no volcano reddens the horizon with its lurid smoke and flame. Ages innumerable have passed away since the glacier flowed down their sides, and left its polished or striated marks on the rocks, to be deciphered by the skill of the geologist; and those hills which once passed through a fiery ordeal, and poured their volcanic floods over the surrounding districts, now form the firmest foundations of the land, and afford quiet, grassy pasturages for the sheep. Our mountains, indeed, possess few or none of those sublime attributes which invest the lofty ranges of other lands with gloom and terror. Their very storms are usually subdued, as if in harmony with their humbler forms. Though they tower to the sky, they seem nearer to the familiar earth; and a large share of the beauty and verdure of the plains do they lift up with them in their rugged arms for the blessing of heaven. Every part of their domains is free and open to the active foot of the wanderer; there are few or no inaccessible precipices or profound abysses to form barriers in his way; he can plant his foot on their highest summits with little expenditure of breath and toil; and a few hours will bring him from the stir and tumult of life in the heart of the populous city to their loneliest and wildest recesses. Well do I love my native hills; for I have spent some of the happiest days of my life in wandering amid their solitudes, following my fancies fearlessly wherever they led me. I have seen them in all seasons, and in all their varied aspects :—in the dim dawn, when, swathed in cold dark clouds, they seemed like "awful countenances veiled," yet speaking in the tongues of a hundred unseen waterfalls; in the still noon-day, when, illumined with sunshine, every cliff and scar on their sides stood out distinctly and prominently against the pure clear sky; at sunset, when, amid the masses of burnished gold that lay piled up in the west—"the glow of fire that burns without consuming"—they seemed like the embers of a universal conflagration; in the holy twilight, when they appeared to melt into the purple beauty of a dream, and the golden summer moon and the soft bright star of eve rose solemnly over their brows, lighting them up with a mystical radiance; and in the lone dark waste of midnight, when from lake and river the long trailing mists crept up their sides without hiding their far-off summits, on which twinkled, like earth-lighted watch-fires, a few uncertain stars. I have gazed upon them in the beauty of summer, when the heather was in full bloom, and for miles they glowed in masses of the loveliest purple; in the changing splendour of autumn, when the deep green of the herbage gave place to the russet hues of the fading flowers, the rich orange of the ferns, and the dark brown of the mosses; and in the dreary depth of winter, when storms during the whole twilight-day howled around them, or when, robed from foot to crown in a garment of the purest snow, they seemed meet approaches to "the great white Throne." In all these aspects they were beautiful, and in all they excited thoughts and emotions which no human language could adequately express.

•Offering such facilities for search, it is not surprising that the vegetable productions of the British mountains should have been thoroughly investigated. Long before Botany became organized as a distinct science, our alpine flora attracted a large share of the attention of scientific men. In the days of Linnæus—stimulated by the enthusiastic impulse communicated by that remarkable man to every department of physical research —a band of devoted botanists undertook the exploration of the Highland mountains; a task by no means so easy then as in this age of steamboats and railroads. The whole of the northern districts encircled by the mighty ramparts of the Grampian range was a *terra incognita*—virtually almost as remote from the civilized regions beyond as the wilds of Labrador. There were no roads, no conveyances, or other means of communication with the south. The adventurous men who first opened up this wild territory to the researches of science were peculiarly adapted for the task of practical scientific pioneers. Endowed with vigorous frames and strong constitutions, they could endure a great amount of privation and fatigue with impunity. The names of Menzies, Lightfoot, Dickson, Stewart, and Don are familiar to every botanist as those of men who contended with innumerable obstacles in the prosecution of their favourite science, then in its feeblest infancy, and popularly regarded with indifference, if not with contempt. The memory of the last-mentioned botanist especially is firmly engrafted in botanical literature, in connexion with his great services in this department. Such was his enthusiastic love of Alpine plants, that he spent whole months at a time collecting them among the gloomy solitudes of the Grampians; his only food a little meal, or a bit of crust moistened in the mountain burn, and his only couch a bed of heather or moss in the shelter of a rock. Before the storms of winter were over, and while the snow still lay far down on the sides of the mountains, he began his wanderings in search of his favourites; and often did he linger on till the last autumn flower withered in the red October sunlight, and the shortening days and scowling heavens warned him of the universal desolation fast approaching. The whole of Western Aberdeenshire and Northern Forfarshire and Perthshire —where the loftiest mountains of Britain have congregated together, storming the sky in every direction with their gigantic peaks, and filling the whole visible scene with themselves and their shadows—was almost as familiar to him as the circumscribed landscape around his native place. Nothing of any interest or importance on these great ranges escaped his eagle eye; and from his numerous visits, and his lengthened sojourn among them, he was enabled to make many interesting discoveries, and to add an unusually large number of species to the flora of Britain. His discoveries were speedily followed up by others, especially by those of Dr. Greville, whose recent death has been one of the greatest losses to botanical science. Professors Graham and Hooker, year after year, conducted their pupils to the summits of the Highland hills; and, not satisfied with a mere cursory visit, they carried tents and provisions with them, and encamped for a week or a fortnight in spots favourable for their investigations. So frequently within the last few years—particularly under the able leadership of Professor Balfour, whose annual class excursions are well known throughout Scotland, and highly prized by all who have the privilege of sharing in them—have the vegetable productions of the principal mountain ranges been investigated, that the most lynx-eyed botanist can now scarcely hope to do more than add a new station for some of the rarer plants; the discovery of a new species being regarded as a very improbable event.

The botanist who takes a comprehensive view of the plants of Great Britain, will find that, excluding exotic species derived from other countries by direct human agency, they may be included in four tolerably distinct groups, which, from their relations to the flora of other parts of Europe, point to a diversified origin. By far the largest portion of our vegetation is composed of forms which are abundant over the whole of Central and Western Europe, and, from their common occurrence on both sides of the German Ocean, have received the name of Germanic plants. In the south-western and southern counties of England, especially where rocks of the cretaceous system prevail, we find a numerous assemblage of plants which are seen nowhere else in the British Isles, and which, from their close relation to the flora of the north-west of France and the Channel Islands, have been denominated plants of the French type. A small but very distinct group of hardy and prolific species is confined to the mountainous districts in the west and south-west of Ireland. These plants, hardly numbering a score, are forms either peculiar to, or abundant in, the peninsula of Spain and Portugal, and especially in Asturias. Lastly, we have the Highland type, which comprehends the species limited to the mountains and their immediate vicinity. This class embraces all the Alpine plants, and contains about a fifteenth of the whole flora of Britain—the number of distinct species amounting to upwards of a hundred. To the most superficial observer, viewed as a whole, they will appear strikingly different from the plants which he is accustomed to see beside his path in the low grounds. The Laplanders and Esquimaux are not more unlike the inhabitants of England and Scotland, than the Alpine flora is unlike that of the plains. The flowers which deck the woods and fields have no representatives in this lofty region. The traveller leaves them one after another behind when he ascends beyond a certain elevation; and though a very few hardy kinds do succeed in climbing to the very summit, they assume strange forms which puzzle the eye, and become dwarfed and stunted by the severer climate and the ungenial soil. All the way up, from a line of altitude varying, according to the character of the mountain range, between two and three thousand feet, you are in the midst of a new floral world, genera and species as unfamiliar as though you had been suddenly and unconsciously spirited away to a foreign country. There are a few isolated islands scattered over the ocean, whose forms of life are unique. St. Helena and the Galapagos Archipelago are such centres of creation, having nothing in common with the nearest mainland. It is the same with the mountain summits in this country that are higher than three thousand feet. They may be compared to islands in an aërial ocean, having a climate and animal and vegetable productions quite distinct from those of the low grounds. Their plants grow in thick masses, covering extensive surfaces with a soft carpet of moss-like foliage, and producing a profusion of blossoms, large in proportion to the size of the leaves, and often of brilliant shades of red, white, and blue; or they creep along the ground in thickly interwoven woody branches, wholly depressed, sending out at intervals a few hard, wrinkled leaves, and very small, faintly-coloured, and inconspicuous flowers. Their roots are usually very woody, or, like those of bulbous plants, wrapped up in membraneous coverings; and their stems are strongly inclined to form buds. They are almost all perennial, the number of annuals being exceedingly small. In all these typical peculiarities, which, it may be remarked, are special adaptations to the unfavourable circumstances in which they are placed, they bear a very close resemblance to the plants of the Polar Zone; and this similarity in the character of the vegetation may be traced from the Arctic regions to the Equator, if we compare, on the mountains of the different zones, the corresponding higher regions, where the isothermal lines are the same, with each other. It must be understood, however, that, except in cases where the plants were originally derived from one centre of distribution, through migration over continuous or closely continuous land, the relationship of Alpine and Arctic vegetation in the Southern Hemisphere, under similar conditions with that of the Northern, is entirely maintained by representative, and not by identical species—the representation, too, being in great part *generic*, and not *specific*.

Strange to say, though so near Europe, the lofty peak of Teneriffe contains on its sides and summit no Alpine flora of a European type. The Retamas of the highest zone are as peculiar to the island as the Euphorbias of the lowest. This absence of northern forms is probably owing to the immense amount of the radiation and the unfavourable hygrometrical conditions of the locality. An equal destitution of Alpine vegetation has been observed on the mountains of Bourbon and Mauritius. On the isolated volcanic peaks of Java, however, though south of the equator, we have plants closely allied to those of the Grampians, while a totally different class of plants clothes the lowlands for thousands of miles around. At a height of 8,000 feet, on the Pangerango mountain, in Java, Mr. Wallace found upwards of forty species, representing European and Alpine genera, and four species actually identical with European species. The Artemisia or southernwood, and the ribwort plantain—the commonest weeds in every British field—occur on this peak at a height of 9,000 feet. Beside them, in the damp shade of the thickets is found the royal cowslip (*Primula imperialis*), which has a tall, stout stem, more than three feet high, with root-leaves eight inches long, and having, instead of a single terminal cluster of cowslip-like flowers, several tiers or whorls, one above another like a Chinese pagoda. This gorgeous cowslip is found nowhere else in the world than on this solitary mountain-summit. On the higher slopes of the Himalayas, and on the tops of the mountains of Central India and Abyssinia, a great many European genera are found, whose existence in such spots Mr. Darwin believes to be owing to the depression of temperature that was so general during the glacial epoch as to allow a few north temperate plants to cross the equator by the most elevated routes of mountain-chains, and to reach the Antarctic regions where they are now found. He believes that the plants on the equatorial summits and the Alpine plants of Europe sprang from a common parentage, and that the modifications which the former have undergone are owing entirely to altered conditions operating during a long period of time. In New Zealand, which is the head-quarters of the Compositæ as well as the ferns, a very remarkable genus of composite plants called *Raoulia* occurs on the sides and summits of the loftier mountains. It numbers twelve distinct species, all of which range from 3,000 to 7,000 feet on Mounts Cook and Doban, and the Nelson and Otago mountains, and form dense, wide-spreading carpets or cushions. The down on the leaves is developed to such an extent as to completely cover them, and almost to conceal the star-like flower-heads. One species, the *R. eximia*, forms gigantic white woolly masses on the ground, and looks at a distance like a flock of sheep grazing on the mountain-side. Indeed, the shepherds are so often deceived by them when folding their charge that the plant has come to be known among the settlers as the "vegetable sheep." This curious genus represents in New Zealand the common cat's-paw or mountain everlasting (*Antennaria dioica*) whose dry white or pink flowers and downy leaves cover our moorlands in myriads; or rather, perhaps, the closely-allied *Gnaphalium supinum*, a tiny cudweed which grows on the extreme summits of the highest Highland mountains. There is one Alpine *Gnaphalium*, peculiar to Greenland and Lapland, called the *G. leontopoides*, or lion's-paw cudweed, whose dense heads are smothered in white silky down. The new Zealand "vegetable sheep" is, therefore, only an extraordinary development of this peculiarity of the tribe even on our own mountains.

There are several curious anomalies in the distribution of Alpine plants, for which no perfectly satisfactory explanation has yet been given. For instance, the genus *Dioscorea* is pre-eminently tropical, being peculiar to the hottest regions of the old and new worlds, the roots of several members being esculent, and used as culinary vegetables, like potatoes. Strange to say, one species of the family, and only one, is found in Europe, the *D. Pyrenaica*, which is an Alpine plant recently discovered at a considerable altitude on the Pyrenees. In like manner, the genus *Pelargonium* is peculiarly African and Australian; and yet a species of it, also an Alpine plant, the *P. Endlicherianum*, has been found on the chain of the Taurus in several stations extending from

Pamphylia to Armenia. The extraordinary genus *Pilostyles*, resembling in miniature the gigantic *Rafflesia* of the Eastern Archipelago—without root, stem, or leaves, and consisting only of a bell-shaped flower sessile on the bark of the tree on which it is a parasite—is peculiar to South America. A species, however, has recently been found in the Alpine regions of Asia Minor, called *P. Hausknechtii;* and though so far out of its proper region, it preserves the peculiar habit of the genus. All the species grow on the bean family; and, true to its native instinct, this Asiatic rover is confined exclusively to a kind of spiny Astragalus. To account for the presence of this South American plant on the mountains of Asia Minor is one of the knottiest points in geographical botany. We can explain in some measure the occurrence of the African *Pelargonium* on the mountains of Taurus on the same grounds that we can account for the remarkable similarity—the almost identity—of the cedar of the Atlas range and the cedar of Lebanon and of the mountains of Taurus. Great changes of surface in very recent geological times have taken place on the African and Asiatic continents, as is proved by many Mediterranean species of fishes being found in the Red Sea and yet not in the Indian Ocean; by the fishes of the salt lakes of Sahara being identical with those of the Gulf of Guinea; and, more extraordinary still, those of the Sea of Galilee with those of the Nile, of the lakes of South-eastern Africa, and the

Zambesi. During some one or other of the great changes of sea and land necessary to produce this remarkable resemblance between the inhabitants of waters now so remote and isolated, the *Pelargonium* may have spread from Africa to Asia Minor. The occurrence of the genus in Australia may be owing to the same cause which produced the resemblance between the marsupial animals, and especially the plants, of Europe during the Eocene epoch, and those of Australia at the present day; a resemblance so striking, that in order to form an idea of the appearance of our country during this geological period we have only to visit our great colony at the Antipodes. But what is the connexion between the sub-Alpine Pilostyles of Asia Minor and the rest of its family in South America? It has been ascertained that the sub-tropical flora of Europe during the *Miocene* epoch is largely American and Japanese. Of the Swiss *Miocene* plants, for instance, no less than 232 species have their nearest allies living in the United States and tropical America, while 108 occur in Asia. In all probability, therefore, the *Miocene* flora of Europe came from America during the Eocene epoch, across the Atlantic, over a great island-continent then existing, which botanists have called "Atlantis," after the ancient legend; and since the Miocene period, this American flora spread from Europe over Asia Minor, Northern Asia, and Japan, in comparatively high latitudes and at considerable elevations, returning to their birthplace by these routes, after having made the circuit of the globe. The present flora of America has descended from these progenitors—the plants of the Eocene and Cretaceous periods of that continent. If we accept this hypothetical Atlantic continent as a scientific fact—established, as it is, by many curious coincidences—it explains to us how there should still be left on the mountains of Asia Minor —like a shell on the shore—a solitary survivor of an ancient American flora identical with the present. The Pilostyles of Asia Minor stands in very much the same relation to the ancient American flora of Europe and Northern Asia, as the one species of myrtle and the one species of laurel now left to us stand to the very large family of myrtles and laurels which spread over Central Europe during the tertiary epoch, and have retreated in these days to the tropical and sub-tropical countries of the two worlds, where they are as numerous as of old.

But in passing from these interesting speculations in general geographical botany to the consideration of our own Alpine flora, a very interesting question arises,—What is the origin of these plants on the British hills? We can hardly suppose them to be indigenous; for they evidently maintain their existence, in the very limited areas to which they are confined, with extreme difficulty, and are comparatively few in number, and poor and meagre in appearance. For these reasons we are fairly entitled to conclude that they are members of specific centres beyond their own area; and these centres must be sought in places where the physical conditions are most favourable for their growth, and where they attain the utmost profusion and luxuriance of which they are constitutionally capable. Now, if we examine the flora of the Lapland and Norwegian mountains, we find that it is not only specifically identical with that of the British Isles, but also that the species of the former are more numerous, and exhibit a greater development of individual forms, than those of the latter, constituting in many places the common continuous vegetation of extensive districts.[1] This fact seems to indicate the Scandinavian mountains as the geographical centres from which we have derived our Alpine plants; and, as might have been expected, allowing this supposition to be true, their gradual migration southwards may be very distinctly traced, like the descent in after ages of the rude Norsemen, by the species left behind on numerous intervening points. On the Faroe Islands, for instance, we have three plants of the Scandinavian type which have stopped short there—viz.

[1] In a collection of fifty-two plants from Baffin's Bay, in lat. 67° and 76° N., made by a friend some years ago, twenty were identical with British species, only somewhat smaller and more stunted. They were gathered during June and July, when the flowers were fully expanded, chiefly on the sea-shore, only three being peculiar to a more elevated locality. The prevailing colour was dark or pale yellow; blue or lilac flowers being comparatively rare. Of the same natural orders seventy-four species occur in Great Britain at an elevation of three thousand feet or upwards.

Saxifraga tricuspidata, *Kœnigia islandica*, and *Ranunculus nivalis*. In the Shetland Islands, the *Arenaria Norwegica*, a common plant on the mountain plateaux of Norway, reaches its southern limits. On the northern shores of the mainland, the beautiful Norwegian primrose appears and ceases. It is called *Primula farinosa*, variety *alpina*, by Norwegian botanists; but it differs in no respect from the *P. Scotica* of Sutherland and Caithness-shire, except in the colour of the flower being paler, the tube a little longer, and the calyx elliptical rather than ovate. A rich assemblage of northern forms is found on the loftiest Highland mountains, distributed apparently from north-east to south-west, in such a manner as to indicate the line of migration. Several species were left behind on the Braemar mountains; while an unusually large proportion is confined to the Breadalbane range, and does not occur further south. Upwards of a score of plants found on the Scottish Alps do not reach the English mountains; while several species are to be met with on Skiddaw and other hills in the north of England which do not extend to the Snowdonian range—Ireland receiving only a few sporadic species. We find the last representatives of this peculiar vegetation on the Alps of Switzerland, at various elevations from 6,000 to 10,000 feet, growing in great luxuriance among a representative flora special in its region,—a few stragglers reaching the Pyrenees in the west, and the Carpathian mountains in the east. We thus find a gradual diminution of the Scandinavian flora as we advance southwards—a convincing proof that it has been diffused in that direction from its original centres of distribution on the elevated ranges of Norway and Lapland. And, regarded from this point of view, Alpine plants may be divided into the *boreal* type, comprehending those species which are confined to the north of Europe, and do not reach farther south than Wales, and the *Alpino-boreal*, which not only extend over the most elevated land in the British Isles, but also occur in abundance at high altitudes on the Swiss Alps and the Pyrenees.

Having thus ascertained the region from which our Alpine vegetation was derived, we have next to account for its transmission. Norway and Britain, at the present day, are widely separated from each other by an extensive sea; and no modes of transportation now in operation are sufficient to account for the diffusion of the peculiar plants of the one country over the mountain ranges of the other, in such a manner as we find them distributed. The problem was quite inexplicable on the supposition formerly entertained, that there has been no striking alteration in the condition of the earth's surface since the present flora of the globe was created, and that the relations of Britain and Norway to each other have always been the same as they are now. It need not be wondered at, therefore, that botanists took refuge from the difficulty in the hypothesis that species have been created indifferently, wherever the conditions were fitted for their growth. But now that we know, from recently ascertained geological facts, that great changes affecting the arrangement of land and water throughout the north of Europe have taken place during the period of the existence of modern vegetation, the key to the mystery has been ascertained.[1]

Attention was first directed to this inquiry by the late lamented Professor E. Forbes, at the meeting of the British Association in 1845; and his views on the subject—supported by the most ample and, I think, conclusive evidence, derived from botanical, geological, and more especially zoological facts—are published at considerable length in the "Memoirs of the Geological Survey." It may seem a superfluous task to direct attention to these views, considering the length of time they have been before the scientific public; but I am persuaded they are not so well known as they ought to be; and to many, a brief popular delineation of them will come with all the interest of novelty.

---

[1] The fishes of the Gulf of Bothnia are identical with those of the Arctic Ocean and White Sea; and yet these fishes occur nowhere on the Norwegian coasts, the only route by which, under the present distribution of land and water, they could have reached the one locality from the other. This circumstance proves that the Baltic and the Arctic Ocean were once connected. It is probable that the plains of Lapland were once under water, and that the Scandinavian peninsula was a group of mountainous islands.

Geological researches have furnished us with two fixed points in time between which this migration of Scandinavian plants to the British hills took place. It must have occurred after the deposition of the London Clay, or the Eocene tertiary epoch; for the organic remains found in that formation belong to a flora very different from, and requiring a far warmer climate than, any now existing on the European continent. And, on the other hand, our great deposits of peat furnish us with conclusive evidence that it must have happened before the epoch usually designated "historical." Between these two periods, geological changes occurred which greatly altered the surface of our islands, and modified their climate and the distribution of their organic forms. From the relics left behind, we learn that a great part of the existing area of Great Britain, especially the lowland plains and valleys, was covered with the waters of a sea which extended over the north and centre of Europe, and was characterised by phenomena nearly identical with those now presenting themselves on the north-east coast of America within the line of summer floating ice. This was the sea of the glacial period—properly so styled—when a condition of climate existed which will account for all the organic phenomena observed in the boulder clays and Pleistocene drifts. In the midst of this sea, the various mountain ranges and isolated hills, which now tower high above the surrounding country, were islands, whose bases and sides were washed by the cold waves and abraded by the passing ice-floes, and whose summits were covered in many places with glaciers, which left their enduring and unmistakeable records on the rocks, and in the moraines at their foot. It was at this period that our now elevated regions received the flora and fauna observed upon them at the present day. Owing to their favourable position in the midst of an ice-covered sea, the means of transport existed in abundance; and the Arctic flora thus brought down, and gradually disseminated over all the islands as far as the sea extended, has ever since been able to maintain its footing, even under the altered climate of our times, according to the general law of climatal influence, through the elevation of the tracts which it inhabits. "This flora would probably differ slightly in different parts of its area, and hence part of the variations now existing between the Alpine floras of Europe. Differences might further result from accidental destruction of the localities of plants scattered sporadically, and from the extinction of forms by various causes during the long period which has elapsed since they first became mountain plants."

There is one remarkable fact which may be noticed in passing, as affording something like circumstantial evidence in favour of this theory. At an elevation of between 3,000 and 4,000 feet on the principal mountain ranges of Scotland, the botanist is astonished to observe the common sea-pink growing among the rocks in the utmost profusion.

It is precisely identical with that which forms so ornamental a feature in the scenery of our sea-coasts; in chemical composition, and in botanical appearance and structure, little or no difference can be detected between specimens gathered in both localities. Nor is it in the Highlands of Scotland alone that the plant is found in such an unusual situation. All over the continent of Europe it occurs on the highest mountains, passing from the coast over extensive tracts of country. It has never been found in the intermediate plains and valleys, except when it has been brought down by mountain streams. This singular circumstance, otherwise inexplicable, would seem strongly to indicate that our mountain chains, as well as those of Northern and Central Europe, were once, as Professor Forbes asserts, islands in the midst of an extensive sea. Plants of sub-Arctic and maritime character would then flourish to the water's edge, some of which would afterwards disappear under altered climatal and physical conditions, leaving the hardiest behind. Another survivor of the ancient maritime flora which once clothed our mountain sides on a level with the glacial waves, is the *Cochlearia Greenlandica*, or scurvy grass, so called from its peculiar medicinal use. Abundant on all our sea-coasts, and never growing inland, it is found in isolated spots at a great elevation on the Highland hills. It may easily be known by its thick tufts, bearing the small white flowers and hot acrid leaves peculiar to the cress tribe. It is so hardy as to defy the severest cold of the Arctic regions, being found by polar navigators in Melville Island, under the snow, at the very farthest limit of vegetation. Farther down, on the sides of our great mountain ranges, we still occasionally observe the *Plantago maritima*, another plant existing nowhere else but on the sea-shore. During the glacial epoch it would flourish in a lower zone than the others, nearer the water's edge, and hence its peculiar altitudinal position at the present day. These three examples, for which no other plausible explanation can be offered, go far to substantiate the theory of the transmission of the Scandinavian flora to our islands, in consequence of the great changes of surface and climate which took place during the glacial epoch.

The plants growing at the present day on the Scottish mountains are thus not only different from those found in the valleys at their base, but they are also much older. They are the surviving relics of what constituted for many ages the sole flora of Europe, when Europe consisted only of islands scattered at distant intervals over a wide waste of waters bristling with icebergs and ice-floes. How suggestive of marvellous reflection is the thought, that these flowers, so fragile that the least rude breath of wind might break them, and so delicate that they fade with the first scorching heat of August, have existed in their lonely and isolated stations on the Highland hills from a time so remote that, in comparison with it, the antiquity of recorded time is but as yesterday; have survived all the vast cosmical changes which elevated them, along with the hills upon which they grew, to the clouds—converted the bed of a mighty ocean into a fertile continent, peopled it with new races of plants and animals, and prepared a scene for the habitation of man! Only a few hundred individual plants of each species—in some instances only a few tufts here and there—are to be found on the different mountains; and yet these little colonies, prevented by barriers of climate and soil from spreading themselves beyond their native spots, have gone on season after season for thousands of ages, renewing their foliage and putting forth their blossoms, though beaten by the storms, scorched by the sunshine, and buried by the Alpine snows, scathless and vigorous while all else was changing around. It is one of the most striking and convincing examples within the whole range of natural history, of the permanency of species!

Our globe may be compared to two enormous snow-capped mountains set base to base at the equator; the Northern Hemisphere representing one, and the Southern the other. The equator is the foot of each; the middle part of both answers to the two temperate zones, north and south; and the opposite summits correspond with the Arctic and Antarctic regions. Thus in each tropical mountain we have an epitome of half of the great earth itself; and all the climates of the world, and all the zones of vegetation, may be felt and seen in passing from its foot to its top in a single day. Altitude is analogous with latitude. To climb a lofty Highland hill is equivalent to undertaking a summer voyage to the Arctic regions; a vertical ascent of 4,000 feet in three hours enabling us to reach a north pole which we could only have attained in as many months by a journey through seventy degrees of latitude. The leading phenomena of the Polar world are presented to us on a small scale within the circumscribed area of the mountain summit. The same specific rocks along which Parry and Ross coasted in the unknown seas of the North, here crop above the surface, and yield by their disintegration the same kind of vegetation. The Alpine hare is common to both; and the ptarmigan, which penetrates in large flocks as far as Melville Island, is often seen flying round the grey rocks of the higher Grampians, and exhibiting its singular changes of plumage from a mottled brown in summer to pure white in winter, so rapidly as to be perceptible from day to day. Although none of the Scotch mountains reach the line of perpetual snow, yet large snowy masses, smoothed and hardened by pressure into the consistence of glacier-ice, not unfrequently lie in shady hollows all the year round, and remind one of the frozen hills of Greenland and Spitzbergen. Sweltering with midsummer heat in the low confined valleys, we are here revived and invigorated by the chill breezes of the Pole. We have thus in our own country, and within short and easy reach of our busiest towns, specimens and exact counterparts of those terrible Arctic fastnesses, to explore which every campaign has been made at the cost of endurance beyond belief—often at the sacrifice of the most noble and valuable lives.

Our Alpine plants may be distributed in three distinct zones of altitude, characterised by Mr. Watson in his admirable "Cybele Britannica" differently from the usual mode. We have first the *super-Arctic* zone, bounded below by the limit of the common heather at an elevation of about 3,000 feet, and defined negatively by the absence, rather than the presence, of particular plants, only two species being peculiar to it in this country. This zone, characterised as that of the herbaceous willow without the heather, occurs only in the Highland provinces, where the highest mountains have their summits considerably above the limits of the heather. We have next, lower down, the *mid-Arctic* zone, lying between the heather line and that of the cross-leaved heath, at about 2,000 feet, characterised by the heather without the heath. This comprehends the highest mountains of England, Wales, and Ireland, and all the great ranges of Scotland, and contains by far the largest proportion of rare and beautiful Alpine plants, being especially rich in Arctic forms. And, lastly, we have the *infer-Arctic* zone, bounded above by the Erica and below by the bracken, and the limits of cultivation at about 1,400 feet. Of course in this zone, which may be characterised as that of the cross-leaved heath without the brake fern, the plants approach more closely to the Lowland type, though containing a large number of species of the true Alpine and Arctic form. These three zones of altitude are distinguished generally by the affinity of their flora to that of the most northern parts of Europe, Siberia, and America, and in a less degree to that of the higher parts of the Swiss Alps, Pyrenees, and Carpathians. We must regard this arrangement, however, though very convenient for general purposes, as somewhat arbitrary and artificial; for Nature is never precise and definite in her lines of demarcation: on the one hand, many Alpine plants growing indiscriminately in all the three zones, and descending in some places even to the sea-shore; while, on the other hand, many common Lowland species come up from the cultivated regions, and grow on the highest summits, although suffering a stunting of their habit from the severer climate. Accidental or local circumstances produce considerable variations in the altitude of the various species. The violent storms which frequently rage in mountain regions sometimes detach fragments of soil, in which several species are rooted, and plant them far down among the productions of the valley; the Alpine streams not only bring down the seeds of Alpine plants, but also, to a certain extent, the cold of the summits, so that their banks will support the species of a severer climate than is natural to the latitude and elevation. On the other hand, deep lakes and other large sheets of water—

as they are less liable to sudden changes than the atmosphere, and preserve a nearly equal temperature all the year round—sensibly mitigate the climate of the mountains in their immediate vicinity, at considerable heights above their surface; hence we not unfrequently find, at an elevation of 2,000 and even 3,000 feet, the plants peculiar to the edge of the water and the lowest declivities blooming in great abundance and luxuriance. On the southern slopes of great ranges which are sheltered from the northern blasts, and more exposed to the light and heat of the sun, the same species are found at a higher altitude than on the northern sides. The range, as well as the character, of the flora is also greatly influenced by the geological construction of the mountains—the number of shady rocks and moist precipices, or comparatively smooth grassy slopes; the direction and nature of the prevailing winds; the frequency of streams and wells; and, above all, by the geographical position of the hills,—whether they form part of an extensive and continuous chain, carrying the general level of the country to a considerable height above the sea-line, and abounding in elevated plateaux and corries, or whether they form conical or isolated peaks rising abruptly from the plains. Considerable allowances must also be made for different latitudes; for though the area of the British Isles is somewhat limited, there is a considerable difference between the temperature of the northern and southern points; so that the isothermal lines of Caithness and Sutherland, at an elevation of 1,300 feet, correspond to those of the summit of Snowdon. The mean annual temperature in the south-west of England is 52°; whereas in the central districts of Scotland it is only 47°, and in the north-east counties as low as 46° or even 45°,—one degree being deducted for inland localities under the same latitude, and one degree for each three hundred feet of elevation above the level of the sea. Attributing their due influence to all these disturbing causes, it will be found, with tolerable regularity and definiteness, that the region occupied by the true Alpine flora extends from an elevation of 2,000 feet to the summits of our highest mountains. This region, as may easily be imagined, is the dreariest and most desolate portion of our country.

Etherealized by the changing splendour of the heavens as the mountain summit appears when surveyed from below, rising up from the huge mound of rock and earth like a radiant flower above its dark foliage, it affords another illustration of the poetic adage, that "'Tis distance lends enchantment to the view." When you actually stand upon it, you find that the reality is very different from the ideal. The clouds that float over it, "those mountains of another element," which looked from the valley like gorgeous fragments of the sun, now appear in their true character as masses of cold, dull vapour; and the mountain peak, deprived of the transforming glow of light, has become one of the dreariest and most desolate spots on which the eye can rest. Not a tuft of grass, not a bush of heather, is to be seen anywhere. The earth, beaten hard by the frequent footsteps of the storm, is bare and leafless as the world on the first morning of creation. Huge fragments of rocks, the monuments of elemental wars and convulsions, rise up here and there, so rugged and distorted that they seem like nightmares petrified; while the ground is frequently covered with cairns of loose hoary stones, which look like the bones which remained unused after nature had built up the great skeleton of the earth, and which she had cast aside in this solitude to blanch and crumble away unseen. When standing there during a misty storm, it requires little effort of imagination to picture yourself a shipwrecked mariner, cast ashore on one of the sublimely barren islands of the Antarctic Ocean. You involuntarily listen to hear the moaning of the waves, and watch for the beating of the foaming surge on the rocks around. The dense writhing mists hurrying up from the profound abysses on every side imprison you within "the narrow circle of their ever-shifting walls," and penetrate every fold of your garments, and your skin itself, becoming a constituent of your blood, and chilling the very marrow of your bones. Around you there is nothing visible save the vague, vacant sea of mist, with the shadowy form of some neighbouring peak looming through it like the genius of the storm; while your ears are deafened by the howling of the wind among the whirling masses of mist, by "the airy tongues that syllable men's names," the roaring of the cataracts, and the other wild sounds of the desert never dumb. And yet, dreary and desolate although the scene usually appears, it has its own periods of beauty, its own days of brightness and cheerfulness. Often in the quiet autumn noon the eye is arrested by the mute appeal of some lovely Alpine flower, sparkling like a lone star in a midnight sky, among the tufted moss and the hoary lichens, and seeming, as it issues from the stony mould, an emanation of the indwelling life, a visible token of the upholding love which pervades the wide universe. If winter and spring in that elevated region be one continued storm, the short summer of a few weeks' duration seems one enchanting festival of light. The life of earth is then born in "dithyrambic joy," blooms and bears fruit under the glowing sunshine, the balmy breezes, and the rich dews of a few days. Scenes of life, interest, and beauty are crowded together with a seeming rapidity as if there were no time to lose. Flowers the fairest and the most fragile expand their exquisitely pencilled blossoms even amid dissolving wreaths of snow, and produce an impression all the more delightful and exhilarating from the consciousness of their short-lived beauty, and the contrast they exhibit to the desolation that immediately preceded.

A large proportion of our Alpine plants are universally diffused, being found in abundance on all the British mountains of sufficient elevation. The Alpine Alchemilla carpets with its satiny leaves the sides of every mountain at an elevation of about a thousand feet; in Braemar it forms the common verdure by the wayside, and mingles with the daisies beside the village houses. The *Sibbaldia procumbens*, somewhat resembling it, is abundant on all the Highland hills, though it does not penetrate farther south; by the roadside on the ascent of the Cairnwall, near Braemar, it is exceedingly common. While the mountain rue (*Thalictrum alpinum*), the white Alpine cerastium, the purple-rayed erigeron, the snowy dryas, the blue veronica, the Alpine saussurea and potentilla, are comparatively common on all the higher ranges of England, Wales, and Scotland. But the most common and abundant of the plants which grow on the Highland mountains are the different species of saxifrage. They are found in cold bleak situations all over the world from the Arctic circle to the equator, and, with the mosses and lichens, form the last efforts of expiring nature which fringe around the limits of eternal snow. A familiar example of the tribe is very frequently cultivated in old-fashioned gardens and rockeries under the name of London pride. Though little prized, on account of its commonness, this plant has a remarkable pedigree. It grows wild on the romantic hills in the south-west of Ireland, from which all the plants that are cultivated in our gardens, and that have escaped from cultivation into woods and waste places, have been originally derived. In that isolated region the London pride is associated with several kinds of heather, with one curious transparent fern, and four or five kinds of lichens and mosses which are found nowhere else in the British Isles, and are eminently typical of southern latitudes. In fact, the same species are again met with on the mountains in the north of Spain; and the theory which botanists have founded upon this remarkable circumstance is, that the south-west of Ireland and the north of Spain were at one period of the earth's history geologically connected, either by a chain of islands or a ridge of hills. Over this continuous land—which we have abundant evidence to prove extended without interruption from the province of Munster beyond the Canary Islands—the gulf-weed, which floats to the west of the Azores, probably indicating the western shore of the submerged continent—flourished a rich and peculiar flora of the true Atlantic type. The intermediate links of the floral chain have been lost by the destruction of the land on which it grew; but on opposite shores of the Bay of Biscay, separated by hundreds of miles, the ends of the chain still exist, amid the wilds of Killarney and the mountain valleys of Asturia. The London pride is, therefore, the oldest plant now growing in the British Isles.

The history of the saxifrages which grow on the Highland hills is scarcely less remarkable—only that they are of Arctic instead of Atlantic origin, and were introduced at a subsequent period into this country. No less than seven different species are found on the Scottish mountains, growing indiscriminately at various altitudes, from the base to the highest summits, on the moist banks of Alpine streams, as well as on bleak exposed rocks where there is hardly a particle of soil to nourish their roots, and over which the wind drives with the force of a hurricane. The rarest of these saxifrages is the *S. cernua*, found nowhere else in Britain than on the extreme top of Ben Lawers, where it seldom flowers, but is kept in existence, propagated from generation to generation by means of viviparous bulbs, in the form of little red grains produced in the axils of the small upper leaves. It resembles the common meadow saxifrage in the shape of its leaves and flower so closely that, though the viviparous bulbs of the one are produced at the junction of the leaves with the stem, and those of the other at the root, Bentham considers it to be merely a starved Alpine variety. Be this as it may, it preserves its peculiar characters unaltered, not only within the very narrow area to which it is confined in Britain, but throughout the whole Arctic circle, where it has a wide range of distribution. So frequently within the last sixty years have specimens been gathered from the station which, unfortunately, every botanist knows well, that only a few individuals are now to be seen at long intervals, and these exceedingly dwarfed and deformed. On no less than twenty-six different occasions I have examined it there, and been grieved to mark the ravages of ruthless collectors. I fear much that, at no distant date, the most interesting member of the British flora will disappear from the only locality known for it south of Norway. After having survived all the storms and vicissitudes of countless ages, historical and geological, to perish at last under the spud of the botanist, were as miserable an anti-climax in its way as the end of the soldier who had gone through all the dangers of the Peninsular war, and was killed by a cab in the streets of London.

The loveliest of the whole tribe is the purple saxifrage, which, fortunately, is as common as it is beautiful. It grows in the barest and bleakest spots on the mountains of England and Wales, as well as those of the Highlands, creeping in dense straggling tufts of hard wiry foliage over the arid soil, profusely covered with large purple blossoms, presenting an appearance somewhat similar to, but much finer than, the common thyme. It makes itself so conspicuous by its brilliancy that it cannot fail to be noticed by every one who ascends the loftier hills in the appropriate season. It is the *avant-courier* of the Alpine plants—the primrose, so to speak, of the mountains—blooming in the blustering days of early April; often opening its rosy blooms in the midst of large masses of snow.

And well is it entitled to lead the bright array of Flora's children, which, following the march of the sun, bloom and fade, one after the other, from April to October, and keep the desolate hills continually garlanded with beauty. It is impossible to imagine anything fairer than a combination of the soft curving lines of the pure unsullied snow, with the purple blooms rising from its cold embrace, and shedding over it the rosy reflected light of their own loveliness. I remember being greatly struck with its beauty several years ago in a lonely corrie far up the sides of Ben Cruachan. That was a little verdant oasis hid amid the surrounding barrenness like a violet among its leaves—one of the sweetest spots that ever filled the soul of a weary, careworn man with yearning for a long repose; walled round and sheltered from the winds by a wild chaos of mountain ridges, animated by the gurgling of many a white Alpine rill descending from the cliffs, carpeted with the softest and mossiest turf, richly embroidered with rare mountain flowers, with a very blaze of purple saxifrage. I saw it on a bright, quiet summer afternoon, when the lights and shades of the setting sun brought out each retiring beauty to the best advantage. It was just such a picture as disposes one to think with wonder of all the petty meannesses and ambitions of conventional life. We feel the insignificance of wealth, and the worthlessness of fame, when brought face to face with the purity and beauty of nature in such a spot. How trifling are the incidents which in such a scene arrest the attention and fix themselves indelibly in the mind, to be recalled long afterwards, perhaps in the crowded city and in the press of business, when the graver matters of every-day life that have intervened are utterly forgotten. High up among the cliffs, round which a line of braided clouds, softer and fairer than snow, clings motionless all day long, rises at intervals the mellow bleat of a lamb, deepening the universal stillness by contrast, and carrying with it wherever it moves the very centre and soul of loneliness. A muir-cock rises suddenly from a grey hillock beside you, showing for a moment his glossy brown plumage and scarlet crest, and then off like the rush of an ascending sky-rocket, with his startling kok-kok-kok sounding fainter and fainter in the distance. Or perhaps a red deer wanders unexpectedly near you, gazes awhile at your motionless figure with large inquiring eyes, and ears erect, and antlers cutting the blue sky like the branches of a tree, until at last, wearied by its stillness, and almost fancying it a vision, you raise your arm and give a shout, when away it flies in a series of swift and graceful bounds through the shadow of a cloud resting upon a neighbouring hill, and transforming it for a moment into the similitude of a pine-forest, over its rocky shoulder, away to some lonely far-off mountain spring, that wells up perhaps where human foot had never trodden.

Speaking of springs, there is no feature in the Alpine scenery more beautiful than the wells and streamlets which make every hill-side bright with their sunny sparkle and musical with their liquid murmurs; and there are no spots so rich in mountain plants as their banks. Trace them to their source, high up above the common things of the world, and they form a crown of joy to the bare granite rocks, diffusing around them beauty and verdure like stars brightening their own rays. A fringe of deeply-green moss clusters round their edges, not creeping and leaning on the rock, but growing erect in thick tufts of fragile and slender stems; clouds of golden confervæ, like the most delicate floss-silk, float in the open centre of clear water, the ripple of which gives motion and quick play of light and shade to their graceful filaments. The Alpine willow-herb bends its tiny head from the brink, to add its rosy reflection to the exquisite harmony of colouring in the depths; the rock veronica forms an outer fringe of the deepest blue; while the little moss campion enlivens the decomposing rocks in the vicinity with a continuous velvet carpeting of the brightest rose-red and the most brilliant green. The indescribable loveliness of this glowing little flower strikes every one who sees it for the first time on the mountains speechless with admiration. Imagine cushions of tufted moss, with all the delicate grace of its foliage miraculously blossoming into myriads of flowers, rosier than the vermeil hue on beauty's cheek, or the cloudlet that lies nearest the setting sun, crowding upon each other so closely that the whole seems an intense floral blush, and you will have some faint idea of its marvellous beauty.[1] We have nothing to compare with it among Lowland flowers. Following the course of the sparkling stream from this enchanted land, it conducts us down the slope of the hill to beds of the mountain avens, decking the dry and stony knolls on either side with its downy procumbent leaves and large white flowers, more adapted, one would suppose, to the shelter of the woods than the bleak exposure of the mountain side. Farther down the declivity, where the stream, now increased in size, scooped out for itself a deep rocky channel, which it fills from side to side in its hours of flood and fury—hours when it is all too terrible to be approached by mortal footsteps—we find the mountain sorrel hanging its clusters of kidney-shaped leaves and greenish rose-tipped blossoms—a grateful salad—from the beetling brows of the rocks ; while, on the drier parts, we observe immense masses of the rose-root stonecrop growing where no other vegetation save the particoloured nebulæ of lichens could exist. This cactus-like plant is furnished with thick fleshy leaves, with few or no evaporating pores ; which enables it to retain the moisture collected by its large, woody, penetrating root, and thus to endure the long-continued droughts of summer, when the stream below is shrunk down to the green gleet of its slippery stones, and the little Naiad weeps her impoverished

---

[1] A sheet of it last summer on one of the Westmoreland mountains measured *five feet across*, and was one solid mass of colour.

urn. Following the stream lower down, we come to a more sheltered and fertile region of the mountain, where pool succeeds pool, clear and deep, in which you can see the fishes lying motionless, or darting away like arrows when your foot shakes the bank or your shadow falls upon the water. There is now a wide level margin of grass on either side, as smooth as a shaven lawn; and meandering through it, little tributary rills trickle into the stream, their marshy channels edged with rare Alpine rushes and carices, and filled with great spongy cushions of red and green mosses, enlivened by the white blossoms of the starry saxifrage. The *S. aizoides* grows everywhere around in large beds richly covered with yellow flowers, dotted with spots of a deeper orange. This lovely species descends to a lower altitude than any of its congeners, and may be called the golden fringe of the richly-embroidered floral mantle with which Nature covers the nakedness of the higher hills. It blooms luxuriantly among a whole host of moorland plants, sufficient to engage the untiring interest of the botanist throughout the long summer day. The curious sundew, a vegetable spider, lies in wait among the red elevated moss tufts, to catch the little black flies in the deadly embrace of its viscid leaves; the bog asphodel stands near, with its sword-shaped leaves and golden helmet, like a sentinel guarding the spot; the grass of Parnassus covers the moist greensward with the bright sparkling of its autumn snow; while the cotton-grass waves on every side its downy plumes in the faintest breeze. Down from this flowery region the stream flows with augmented volume, bickering over the shingle with a gay poppling sound, and leaving creamy wreaths of winking foam between the moss-grown stones that protrude from its bed. It laves the roots of the crimson heather and the palmy leaves of the lady-fern. The sunbeams gleam upon its open face with "messages from the heavens;" the rainbow arches its waterfalls; the panting lamb comes to cool its parched tongue in its limpid waters; the lean blue heron, with head and bill sunk on its breast, stands motionless in its shallows watching for minnows all the long dull afternoon, while the dusky ousel flits from stone to stone in all the fearless play of its happy life. Hurrying swiftly through the brown heathy wastes that clothe the lower slopes, it lingers a while where the trembling aspen and the twinkling birch and the rugged alder weave their leafy canopy over it, freckling its bustling waves with ever-varying scintillations of light and shade; pauses to water the crofter's meadow and cornfield, and to supply the wants of a cluster of rude moss-grown huts on its banks, which look as if they had grown naturally out of the soil; and then, through a beach of snow-white pebbles, it mingles its fretting waters in the blue, profound peace of the loch. Such is the bright and varied course of the Alpine stream, with its floral fringe; and from its fountain to its fall it is one continuous many-linked chain of beauty—an epic of Nature, full of the richest images and the most suggestive poetry.

Very few of the true Alpine plants grow on the actual summits of the Highland hills; and this circumstance appears to be due not so much to the cold—for the same plants are most abundant and most luxuriant throughout the whole Polar zone, where the mean annual temperature is far below the freezing point, whereas that of the Highland summits is 3° or 4° above that point—but to their want of shelter from the prevailing storms, and the generally unfavourable geological structure of the spots. The highest point of Ben Nevis, for instance, is so thickly macadamized with large masses of dry red granite, that there is hardly room for the tiniest wild flower to strike root in the soil. It looks like the battle-ground of the Titans, or a gigantic heap of scoriæ cast out from Vulcan's furnace. It is impossible to imagine, even in the Polar regions, any spot more barren and leafless. The plants of the super-Arctic and mid-Arctic zones, which should be found there owing to its height, are therefore obliged to accommodate themselves in the infer-Arctic zone, where the necessary conditions of soil and moisture exist. One of the two plants characteristic of the highest zone—viz. the *Saxifraga rivularis*—occurs on the hill, but considerably below its normal limits. It grows at an altitude of 3,000 feet, in a spot irrigated, while the plant is in flower, by water trickling from the melting snow above. The summit of Ben-y-gloe, rising to a height of 3,900 feet in the north-east corner of Perthshire, is also covered with enormous piles of snowy gneiss—like the foundation of a ruined city, in some places ground into powder by the disintegrating effects of the weather, and in others occurring in the shape of large blocks thrown loosely above each other, so sharp and angular that it is one of the most difficult and fatiguing tasks imaginable to scramble over the ridge to the cairn which crowns the highest point. When surveyed from below, the peak has a singularly bald appearance, scarred and riven by numberless landslips, and the dried-up beds of torrents, and scalped by the fury of frequent storms; and a nearer inspection proves it to be as desolate and leafless as the sands of Sahara. On the top of Ben-Mac-Dhui, though very broad and massive, as beseems a mountain covering a superficial basis of nearly forty miles in extent, the only flowering plants which occur are, strange to say, those which are found in profusion even at the lowest limits of Alpine vegetation on the English hills. The last time I visited it I observed only seven flowering plants near the cairn on the summit, most of which were sedges and grasses. The mossy campion, however, amply compensated me for the absence of the other Alpines by the abundance and brilliancy of its rosy flowers.

The same remarks apply to nearly all the Highland hills. There are only five plants which—though sometimes descending to lower altitudes, one or two of them even to the level of the sea-shore on the hills fronting the coast in the north-west of Scotland—are invariably found on the summits of all the ranges that are more than 3,000 feet high. These plants are the mossy campion, the Arctic willow, the procumbent sibbaldia, the little dusky-brown gnaphalium, and the curious cherleria or mossy cyphel. This last little plant forms an anomaly in the distribution of our Alpine flora. It is very abundant in the subnival region of the Swiss Alps, growing on the larger groups of mountains, from an altitude of 8,000 to 15,000 feet. It forms one of the most conspicuous of the forty plants found on the far-famed "Jardin de la Mer de Glace" at Chamouni, described in Murray's Handbook as "an oasis in the desert, an island in the ice, a rock which is covered with a beautiful herbage, and enamelled in August with flowers. This is the Jardin of this palace of nature; and nothing can exceed the beauty of such a spot, amidst the overwhelming sublimity of the sur-rounding objects—the Aiguille of Charmoz, Bletier, and the Géant," &c. This highly-coloured description is, however, a mere euphemism, for in reality the so-called garden is only a rock protruding out of the glacier, and covered principally with lichens and plants whose dull, insignificant appearance would not attract the least notice elsewhere. Although not very rare on the highest Scottish mountains, the cherleria does not extend farther north—thus offering a very striking exception to the usual derivation of our mountain flora. It may either have emigrated northwards from the Alps during the glacial epoch, or it may be regarded as a sporadic species, depending upon local conditions for its maintenance. From its peculiar and hardy appearance, we would almost hazard the opinion that it is older than any of the other Alpine plants, that it existed on the British hills before the migration of the Scandinavian flora, and that the Breadalbane mountains form its original centre, from which it has been distributed southwards over the Swiss Alps. The last inference is warranted by its extraordinary luxuriance on Ben Lawers. It has nothing to boast of in the shape of flowers, the sharpest eyes being hardly able to detect the minute greenish petals and stamens among the tufted moss-like foliage. It is impossible to convey the impression of special adaptation which one glance at the plant, in its bare and sterile habitat, cannot fail to produce. Its long, tough, woody root penetrates deeply the stony soil, so that it is with difficulty a specimen can be detached; and so hardy is its nature that it flourishes green and luxuriant under the chilling pressure of huge masses of snow, and under the unmitigated glare of the scorching summer sun.

Of all the British mountains, Ben Lawers is the richest in rare and interesting Alpine species. This hill, which may be called the Mecca of the botanist, as every neophyte who aspires to the honours of his science must pay a visit to its rugged cliffs, occupies very nearly the centre of Scotland. It rises in a pyramidal form from the north shore of Loch Tay, upwards of 4,000 feet above the level of the sea, and commands from its summit, on a clear day, an uninterrupted view unparalleled in the British islands for variety, sublimity, and extent. Though separated from the surrounding mountains by two torrents which flow through deep depressions on its eastern and western sides, it forms with them an immense continuous range, upwards of forty miles in length, ten in breadth, and of an average altitude of 3,000 feet. On this lofty plateau, known as the Breadalbane chain, which is the most uniformly and extensively elevated land in Britain, the different peaks of Maelghyrdy, Craigcalleach, Ben Lawers, &c., repose like a conclave of mighty giants, imparting a serrated appearance to the range indescribably wild and savage when wreathed with mist or cloud. The whole of this vast region is composed almost entirely of micaceous schist, interspersed here and there with veins of quartz, and containing not unfrequently those dark-brown crystals called garnets, which greatly enhance the sparkling lustre of the mica. This rock, it may be remarked, embraces within its course the finest and most celebrated scenery in the Highlands, and rises, besides the Breadalbane peaks, into such distinguished summits as Ben Voirlich, Ben Ledi, Ben Venue, Ben Lomond, and all the bold serrated ridges of Argyleshire and Inverness-shire. It is of a very soft and friable nature, and is easily weathered, forming on its surface a deep layer of rich soil, admirably adapted to the wants of an Alpine or Arctic vegetation. Being the prevailing formation in the Norwegian and Lapland mountains, as well as in the Arctic regions, it is obvious that the Scandinavian plants which emigrated southwards would find, wherever this rock cropped out sufficiently high above the surrounding surface, peculiarly favourable conditions for their growth. Hence on all the micaceous rocks in this country, and even in the Swiss Alps, we find a greater variety and a richer luxuriance of Scandinavian forms than on any other geological formation. We are particularly struck with this when we compare the rich and varied Alpine vegetation of the Breadalbane mica schists with the generally meagre and stunted vegetation of the Braemar and Ben Nevis granites. The unusual fertility of the Breadalbane range must also be ascribed to geographical position, highly advantageous in a meteorological point of view. The south-west winds, which come loaded with moisture from the Atlantic, meet with this great ridge running along the west of Perthshire, high above the other ranges, and, rushing up its cooler sides, condense their vapours, disengage their latent heat, and produce that mild climate, with almost continual rain or drizzling mist, in which Alpine plants delight during the period of growth; whereas to the Aberdeenshire mountains the same winds come deprived of their moisture, and bring dry, cold weather. The common species of plants which are found on every hill of sufficient altitude in Britain, and which constitute their sole Alpine flora, are not only more abundant in individual forms on the Breadalbane mountains, but also attain more luxuriant proportions, so that they give a rich and beautiful appearance to the higher ranges in the glowing summer months, while, as previously intimated, an unusually large proportion of plants is exclusively restricted to this chain. Nor is it merely in rare phanerogamous vegetation that these mountains are rich; they also possess a singularly varied and peculiar cryptogamic flora, several species of which are found nowhere else. Most of these plants may be found collected on the single peak of Ben Lawers; and a botanist cannot spend a week more profitably and pleasantly than in exploring the huge sides and broad double summit of this hill. Every step leads to a botanical surprise, and almost every plant is either altogether new, or so rare and unfamiliar as to excite a thrill of gratification. If he has never before investigated Alpine vegetation, and if he be at all an enthusiast in his pursuit, he will experience in the collection of these novelties and rarities some of the happiest moments in his life, —moments worth years of artificial excitement, banishing every sense of weariness and fatigue, and rendering, by the elevation of mind they produce, his perceptions of beauty in the scenery around more acute and delightful. These moments soon pass away, but they cease like the bubbling of a fountain, which leaves the waters purer for the momentary influence which had passed through them,—not like too many worldly joys, which ebb like an unnatural tide, and leave behind only loathsomeness and disgust.

In the crevices of the highest rocks may be observed a curious lichen, called *Verrucaria Hookeri*, spreading over the blackened and hardened turf in white turgid scales, which is quite different from any other lichen with which we are acquainted, and seems to be a special creation found nowhere else in the world. Curiously enough, there is associated with it a moss also peculiar to the spot, the *Gymnostomum cæspititium*, which grows in dense brownish-green tufts, with numerous glossy capsules nestling among the leaves. The extreme rarity and isolation of these plants would almost warrant the inference, either that they are new creations which have not yet had time to secure possession of a wider extent of surface, or rather, perhaps, that they are aged plants, survivors of the original cryptogamic flora of the soil during the more recent geological epochs, which have lived their appointed cycle of life, and, yielding to the universal law of death, are about to disappear for ever. On the highest ridge of the mountain occurs, among the *débris* of rocks, the *Draba rupestris*, a very small, insignificant-looking plant, but important as being one of the most Arctic and Alpine plants in Scotland. It is only found here and in one locality in Sutherlandshire, and is unknown on the Continent of Europe. Passing down from the cairn that crowns the highest point of Ben Lawers, along the north-western shoulder of the hill, we are soon brought to a stand by several lofty precipices. Descending one of these, we come to a small corrie; and here, upwards of 3,000 feet above the level of the sea, we are fairly bewildered with the beauty, the variety, and the luxuriance of the Alpine plants which bloom on every side. All the ordinary species are here congregated in lavish profusion, protected by immense shaggy beds of rare Alpine mosses, and nourished by the incessant dripping from the rocks overhead. We observe among them a few dense tufts of the Alpine sandwort (*Alsine rubella*), and instantly we are down on our knees in the swamp to gather it, for one brief moment oblivious of the whole universe besides. My prize has certainly little to recommend it; for beauty it can scarcely be said to possess, the chickweed of our gardens, to which it is closely allied, having fully as pretty a flower; but it is remarkable for that which gives value to the diamond—its exceeding rarity—only one other station for it being known in Britain, viz. the exposed cliffs of Ben Hope in Sutherlandshire. It belongs eminently to the boreal or Arctic type of vegetation, penetrating very far north, but reaching its southern limit on Ben Lawers. Scarcely has my enthusiasm had time to cool, when it is raised to a higher pitch, by seeing, in a cleft of the rock, the most celebrated of all our mountain flowers—the tiny *Gentiana nivalis*, or snowy gentian. With immeasurable thankfulness, and with a reverential and delicate touch, I pluck from the tiny clumps two specimens for myself, and two for favoured friends—no more; for the genuine botanist has too great a regard for these interesting remnants of an almost extinct race—these little Aztecs of the flower world, which cling so tenaciously to Flora's skirts—to exterminate them ruthlessly by taking more than he needs. If, humanly speaking, they are so precious in the eyes of their Creator, that He has taken such wonderful care to perpetuate them in these bleak spots, they ought surely to be invested with something of a sacred character in our sight. What appeals so powerfully to the protection of man in the helpless form of the infant, ought to affect us in similar, though of course lesser degree, in the tenderness and fragility of these rare plants. The snowy gentian is the smallest of the Alpine flowers, usually averaging from half an inch to an inch in height, with a very minute blossom, forming a mere edge of deep blue, tipping the long calyx. Another station besides the Ben Lawers one has been found in the Caenlochan mountains, at the head of Glen Isla, where a porphyritic granite, rich in felspar, associated with a dark syenite, abounding in hornblende, is the prevailing rock. The Alps of Switzerland, however, seem to be the chosen haunt of this and all the rest of the gentian tribe.* There it grows in profusion among a lovely sisterhood of gentians, imparting a blue, deep as that of the sky above, to the higher pasturages, and often hides its head on the dizzy ledges of tremendous precipices. In ascending the lofty peaks of the Jungfrau and Monte Rosa, the guides not unfrequently resort to the innocent artifice of endeavouring to interest the traveller in its beauty, to distract his attention from the fearful abysses which the giddy path overhangs.

There is one flower found in Ben Lawers which alone is worth all the fatigue of the ascent. This is the Alpine forget-me-not (*Myosotis alpestris*). It is far lovelier than its sister of the valleys—the well-known flower of friendship and poetry—its flowers being larger, more numerous, and closely set, forming a dense coronet or clustered head, that looks like a carcanet of rich turquoises. It does not grow beside running brooks, or in marshy spots, like its lowland congener, but high up on the dizzy ledges of almost inaccessible cliffs, where no one but the prying naturalist would look for floral beauty. Though somewhat abundant on the Swiss Alps, in Britain it is confined to the Breadalbane mountains, where it does not occur lower down than 3,000 feet. On Ben Lawers it is especially abundant and luxuriant, crowning with a garland of large blue tufts the precipitous crags which jut out from the western side of the hill. Fortunately for the preservation of the plant, it is a hazardous undertaking to gather it there, for the rocks are from 300 to 400 feet in perpendicular height, and one escapes from their ledges to a secure standing-place with much the same feelings that a man gets out of reach of a mortar just about to explode. In that elevated spot the summer is far advanced before it ventures to put forth its delicate flowers, so that it escapes the howling winds and the tempestuous mists, and blooms in a calm and serene atmosphere. The perfume which it exhales is very volatile, being sometimes almost imperceptible, and at other times very strong, and suggestive of the honey smell of the clover fields left far below. This is almost the only British Alpine plant possessed of fragrance; whereas, on the Swiss Alps, the majority of species are odoriferous,—a circumstance which adds largely to the inspiring influence of a ramble on those stupendous hills. The absence of scented species on our mountains seems to be owing to the dark cloudy atmosphere which almost always broods over them; while their presence in such profusion on the Alps is, on the other hand, due to the cloudless skies and the bright sunshine peculiar to the south, as well as to the diminished pressure of the atmosphere; for the most fragrant kinds seldom prosper below a certain elevation, and when cultivated in gardens become nearly scentless. There is no plant which recalls more forcibly the beautiful though hackneyed lines of Gray than the Alpine forget-me-not. But is it really true that it blushes unseen, and wastes its fragrance on the desert air? Who are we, that we should arrogate to ourselves the right to call any existence vain and wasted that is wholly beyond our use, and removed from our admiration? When shall we learn the humbling truth, constantly preached to us, that nature has not yet passed under our dominion, and that the smallest wild flower does not bloom for man, or any other creature, as its primary object. We have seen how little the admiration of man is regarded by nature, in the boundless prodigality with which she pours out her treasures in the loneliest and most desolate spots, remote from human habitations, and rarely, if ever, visited by human foot. There are many beautiful scenes left far off by themselves among the solitudes of the mountains, where, "unseen and unknown to all human beings, living nature fails not, from the glad morn to the silent eve, to call up all those sublime pageants of daily recurrence which show forth the Creator's unchangeable glory, in her ever-changing loveliness; where the sunrise, unnoticed, clothes the mountains with regal robes of crimson and gold, and the red twilight, unadmired, paints them in hues soft as those which pass over the cheek of the dying; where grateful flowers, ungathered, breathe forth their odours like the incense of a silent prayer, while answering dews descend, untainted, from the skies; where storms unfeared come down in all their terror, and the unheard winds make a ceaseless wailing music over the lonely heights." And are we to think that all these beauties and wonders of creation are lost, because no mortal is at hand to look on them with his cold eye and thankless heart? No! better to suppose that purer and holier eyes than ours are for ever keeping watch in grateful admiration over the minutest flower, as over the remotest star, than to believe that the works of the Creator are ever without some one of His created beings to adore His majesty in their perfection.

The Aberdeenshire mountains, from their great elevation and geographical position, lying in one of the directions taken by the Scandinavian flora in its descent to southern latitudes, exhibit a large proportion of Alpine forms, which might have been still larger were it not for unfavourable geological and climatal conditions. They possess, in great luxuriance, on the sides and summits of their highest peaks, no less than three species of shrubby lemon-coloured lichens highly peculiar to Iceland and Lapland, and found nowhere else in this country. The restriction of these cryptogams to so narrow a corner of our island—considering the facility with which their light, invisible spores may be disseminated by winds and waves, and their capacity of enduring the utmost extremes of temperature—can only be explained by the supposition that the Cairngorm mountains first intercepted and retained them. Of phanerogamous plants, two at least are confined to this district. Of these, the *Mulgedium alpinum*—a large, coarse plant of the thistle tribe, with erect stems from two to three feet high, producing deep blue florets late in summer—

grows in moist, rocky situations in Northern and Arctic Europe and Asia; but in this country is restricted to the Loch-na-gar and Clova mountains, where it is rapidly disappearing. I gathered it several years ago in a locality where I believe it is now extinct,—the ledge of a sloping and rugged precipice on the north side of Ben-Muich-Dhu, down which a stream, rising in the upper ranges of the hill, falls in a succession of cascades for nearly 3,000 feet into the waters of Loch Avon. The other is the *Sonchus alpinus*, or Alpine sow-thistle, an equally coarse plant. It is found on the same cliffs of Loch-na-gar on which the *Mulgedium* grows. On the rocks overhanging a deep ravine, by which there is an ascent—though very laborious —to the summit, may be found *Saxifraga rivularis* and *Phleum alpinum;* while the rare *Lycopodium annotinum, Cornus suecica,* and *Drosera anglica* may be gathered at their base in moist soil.

On the Braemar mountains another Alpine plant of deeply interesting character is found. The *Astragalus alpinus*—a species of vetch—crowns the summit of Craigindal, a hill about 3,000 feet high, in the vicinity of Ben Avon and Ben-na-bourd. It is confined almost exclusively to this neighbourhood, and is found there in two or three localities at considerable distances from each other, but characterised by the same geological formation, viz. a very pure, compact felspar. These mountains form the most southern limit of this plant. Tracing the Grampian chain for twenty or thirty miles south-east, until it forms the Clova group of hills, we find collected in that narrow space two other plants, each of which is restricted in its range to rocks of the same specific character, and therefore comprised within a very limited area. One of these, the *Oxytropis campestris*—also a species of vetch, with pale yellow flower tinged with purple—is known by reputation, if not by sight, as one of the rarest of British plants, and therefore one of the most desirable acquisitions to the herbarium. Common on the mountain pastures and Alpine rocks in the Arctic regions of Europe, America, and Siberia, it is confined in Britain to one cliff in Clova, severed from the surrounding precipices by two deep fissures, apparently the result of extensive atmospheric disintegration. This cliff is composed of micaceous schist, peculiarly rich in mica, though of a dark smoky colour; and being of a soft and friable nature, easily decomposed by the weather, forms a loose, deep, and very fertile soil. The other plant alluded to, viz. the *Lychnis alpina*, is also confined to a few isolated localities in the same range. It grows sparingly on the rocky table-land —about half an acre in extent—which crowns the summit of a hill called Little Gilrannoch, equidistant between Glen Isla and Glen Dole. It is intimately connected with the lithological character of its habitat, for in several places on this plateau it springs from little crevices where there is hardly a particle of soil to nourish its roots; and its range of distribution extends only as far as the rock preserves its mineral character unchanged. This rock, which differs from the prevailing strata of the district, and from those in its immediate neighbourhood, is composed of compound felspar, very hard, and capable of resisting disintegration. In some places it is smooth and bare, like a pavement, and in others extremely corrugated and vitrified, as if it had undergone the action of fire. Though not found elsewhere in this country, the Alpine *Lychnis* has an extensive geographical range, being an Alpino-boreal plant, occurring both in Scandinavia and the Swiss Alps and Pyrenees.

Caenlochan stands next, perhaps, to Ben Lawers in the number and interest of its Alpine rarities. On the summit of this range, close beside the bridle-path which winds over the heights from Glen Isla to Braemar, an immense quantity of the Highland azalea (*Azalea procumbens*) grows among the shrubby tufts of the crowberry; and when in the full beauty of its crimson bloom, about the beginning of August, it is a sight which many besides the botanist would go far to see. It is the only plant on the Highland mountains that reminds us of the rhododendrons which form the floral glory of the Swiss Alps, and especially of the Sikkim Himalayas. The stupendous cliffs at the sources of the Isla, formed of friable micaceous schist, and irrigated by innumerable rills, trickling from the melting snow above, are fringed with exceedingly rich tufts of *Saussurea, Erigeron, Sibbaldia, Saxifraga nivalis*, and whitened everywhere by myriads of

Dryas octopetala and Alpine *Cerastium*. The scenery of this spot is truly magnificent. Huge mural precipices, between two and three thousand feet high, extend several miles on either side of a glen so oppressively narrow that it is quite possible to throw a stone from one side to the other. Dark clouds, like the shadows of old mountains passed away, continually float hither and thither in the vacant air, or become entangled in the rocks, increasing the gloom and mysterious awfulness of the gulf, from which the mingled sounds of many torrents, coursing far below, rise up at intervals like the groans of tortured spirits. A forest of dwarfed and stunted larches, planted as a cover for the deer, scrambles up the sides of the precipices for a short distance, their ranks sadly thinned by the numerous landslips and avalanches from the heights above. This region is seldom frequented by tourists, or even by botanists, as it lies far away from the ordinary routes, and requires a special visit. The late Professor Graham and the present accomplished Professor of Botany in the Edinburgh University once spent, I believe, a fortnight in the shieling of Caenlochan, a lonely shepherd's hut at the foot of the range, built in the most primitive manner and with the rudest materials. They gathered rich spoils of Alpine plants in their daily wanderings among the hills, and so thoroughly indoctrinated the shepherds and gamekeepers about the place in the nature of their pursuits, that they have all a knowledge of, and a sympathy with, the vasculum and herbarium, rare even in less secluded districts, though the schoolmaster is everywhere abroad. Every one of them knows the "Gimtion" (*Gentiana nivalis*) and the "Lechnis amēna" (*Lychnis alpina*), as they call them, as well as they know a grouse or sheep, and is proud at any time, without fee or reward, to conduct "botanisses" to the spots where these rarities are found.

In the northern extremity of Perthshire, between Loch Rannoch and Loch Erricht, on the north-eastern brow of the mountain called the Sow of Atholl, is the well-known station for the very rare *Menziesia cœrulea*, a species of heath distinguished by its large blue bells. This treeless waste of elevated moorland, characterised by Maculloch as one of the most desolate regions in Europe, forgotten by nature, without a trace or a recollection of human life, once formed the site of the great Caledonian forest, which, in all probability, sheltered in its moist and shady recesses plants found nowhere else in Britain, and peculiar to the swampy forests of Norway and Lapland. Of this hyperborean vegetation, the beautiful *Menziesia* and the *Rubus arcticus* are now the sole surviving relics. They strikingly illustrate the influence of man in extirpating or limiting the distribution of plants, by levelling forests, draining marshes, and thus rendering a particular region unsuitable to the vegetation of an excessive climate, by introducing a more equable temperature, greater warmth in winter and greater cold in summer, than formerly prevailed. To the general naturalist this is one of the most interesting districts in Britain. About nine miles from Kinloch Rannoch, on the south side of the loch, there is a thick, dark pine-forest known as the Black Wood, which is also a relic of the great Caledonian forest; many of its trees being of great age, and so large as to require the outstretched arms of two men to span them. In the damp air of this forest, where there is an abundant supply of vegetable food in all stages of decay—favoured by the intense heat of summer and the long period of winter torpor—an astonishingly large number of subalpine insects occur, which are unknown elsewhere. It is, in fact, the paradise of the entomologist, for though the species are rare, the number of individuals is unusually large. Many of them are of considerable size, and possess very attractive colouring; while others exhibit curious habits and modes of development. The *Formica congerens* builds its huge anthills of pine-needles here as in Norway. One of the most abundant insects in the place is the Longicorn beetle (*Astinomus ædilis*), which is known in Sweden, and, strange to say, in Rannoch also, as "the timberman," on account of its frequenting the timber-cutting yards, and even the doorposts of the houses. Its horns are prodigiously long, about four times the length of its body, and remind one more of tropical insects than any similar development that occurs in this country. *Trichius fasciatus*, known to the villagers as the "bee-beetle," from the resemblance of the velvety black bands on its yellow downy body to those of the common humble-bee, is also frequent in the neighbourhood. In short, upwards of a score of insects peculiar to the neighbourhood are essentially boreal forms. The parallelism between them and the insects of Norway and Sweden is of the closest character, and is thus a singular confirmation of the evidence afforded by some of the plants of the district that, in this corner of Britain, we have, in the relics of the Caledonian forest, the remains of a Scandinavian flora and fauna that once spread over the whole country.

Although neither tree nor shrub is capable of existing on the mountain summits, we find several representatives there of the lowland forests. The Arctic willow (*Salix herbacea*) occurs on all the ridges, creeping along the mossy ground for a few inches, and covering it with its rigid shoots and small round leaves. It is a curious circumstance, that a regular sequence of diminishing forms of the willow tribe may be traced in an ascending line, from the stately "siller saugh wi' downie buds," that so appropriately fringes the banks of the lowland river, up to the diminutive species that scarcely rises above the ground on the tops of the Highland hills. The dwarf birch, also, not unfrequently occurs in sheltered situations on the Grampians, among fragments of rocks thickly carpeted with the snowy tufts of the reindeer moss. It is a beautiful miniature of its graceful sister, the queen of Scottish woods, the whole tree—roots, trunk, branches, leaves, flowers, and fruit—being easily gummed on a sheet of common note-paper; and yet it stands for all that the Esquimaux and Laplanders know of growing timber. In the Arctic plains the members of the highest botanical families are entirely superseded by the lowest and least organized plants. Lichens and mosses are there not only more important economically, but have greater influence in affecting the appearance of the scenery, than even willows and birches.

Of ferns there are several very interesting species on the Highland mountains. A peculiar form occurs in sheltered places on most of the higher summits, which for a long time was supposed to be a variety of the common lady-fern. It is now ascertained to be a distinct species, and is called *Polypodium alpestre;* its cluster of spores being naked and destitute of a covering in all the stages of growth. It is especially abundant on Loch-na-gar and the Cairngorm range, where it was discovered several years ago by Mr. Backhouse of York. On rocky slopes, at a height of about 2,000 feet, occurs the Alpine holly fern (*Polystichum Lonchitis*), which is peculiarly adapted to its rigorous climate by its slow rigid habit of growth, and the persistency of its old fronds. On Ben Lawers it is very abundant. The *Woodsia hyperborea* grows in small compact tufts on the ledges of almost inaccessible precipices. It is confined almost exclusively to the Breadalbane mountains, where it is found very sparingly indeed. But the rarest and most interesting of all the Alpine ferns is the *Cystopteris montana*, a large, handsome, much-divided species, bearing a considerable resemblance to the *Polypodium calcareum*. It appears at the beginning of June, and fades early in August. It does not grow in crevices of rocks, like its congeners, but on the Alpine turf at a height of about 3,000 feet. On Ben Lawers I once observed an extensive patch of it, containing thousands of specimens; but when I next visited the spot, the turf had been stripped off and the plant extirpated—not a vestige of it to be seen. It is fortunately, however, abundant on the wild, almost unknown, mountain plateaux which stretch from the head of Loch Tay to Loch Lomond—such as Benteskerny, Mael-nan-tarmonach, Maelghyrdy, Corry Dhuclair, &c. I gathered some fine specimens of it in a ravine while crossing the Wengern Alp, in Switzerland, some years ago; and subsequently in the tremendous defile of the Nâerodal, at the head of the Sogne Fjord in Norway. Its original centre of distribution seems to be the Rocky Mountains of America, for there it occurs in the utmost luxuriance and profusion.

It would be improper not to notice very briefly the rich and varied cryptogamic vegetation which clothes the highest summits, and spreads, more like an exudation of the rocks than the produce of the soil, over spots where no flowering plant could possibly exist. This vegetation is permanent, and is not affected by the changes of the seasons: it may, therefore, be collected at any time, from January to December. It is almost unnecessary to say, that the Alpine mosses and lichens are as peculiar and distinct in their character from those of the valleys as the Alpine flowers themselves. They are all eminently Arctic; and, though they occur very sparingly in scattered patches on the extreme summits of the Highland hills, they are the common familiar vegetation of the Lapland and Iceland plains, and cover Greenland and Melville Island with the only verdure they possess. Some of them are very lovely: as, for instance, the saffron *Solorina*, which spreads over the bare earth, on the highest and most exposed ridges, its rich rosettes of vivid green above and brilliant orange below; the daisy-flowered cup lichen, with its filigreed yellow stems, and large scarlet knobs; and the geographical lichen, which enamels all the stones and rocks with its bright black and primrose-coloured mosaic. Some are useful in the arts, as the Iceland moss, which occurs on all the hills, from an elevation of 2,000 feet, and becomes more luxuriant the higher we ascend. On some mountains it is so abundant that a supply sufficiently large to diet, medicinally, all the consumptive patients in Scotland could be gathered in a few hours. A few lichens and mosses, such as Hooker's *Verrucaria*, and Haller's *Hypnum*, are interesting to the botanist, on account of their extreme rarity and isolation. Some are interesting on account of their associations, as the *Parmelia Fahlunensis*, which was first observed on the dreary rocks and heaps of ore and *débris* near the copper mines of Fahlun, in Sweden—a district so excessively barren that even lichens in general refuse to vegetate there, yet inexpressibly dear to the great Linnæus, because there he wooed and won the beautiful daughter of the learned physician Moræus. And the curious tribe of the *Gyrophoras* or *Tripe de Roche* lichens, looking like pieces of charred parchment, so exceedingly abundant on all the rocks, will painfully recall the fearful hardships and sufferings of Sir John Franklin and his party in the Arctic regions. It is a strange circumstance, by the way, that most of the lichens and mosses of the Highland summits are dark-coloured, as if scorched by the fierce unmitigated glare of the sunlight. This gloomy Plutonian vegetation gives a very singular appearance to the scenery, especially to the top of Ben Nevis, where almost every stone and rock is blackened by large masses of *Andreas*, *Gyrophoras*, and *Parmelias*.

The most marked and characteristic of all the cryptogamic plants which affect the mountain summits is the woolly-fringe moss. This plant grows in the utmost profusion, frequently acres in extent, rounding the angular shoulders of the hills with a padding of the softest upholstery work of nature ; for which considerate service the botanist, who has previously toiled up painfully amid endless heaps of loose stones, is exceedingly grateful. Growing in such abundance, far above the line where the higher social plants disappear, it seems a wise provision for the protection of the exposed sides and summits of the hills from the abrading effects of the storm. Snow-wreaths lie cushioned upon these mossy plateaux in midsummer, and soak them through with their everlasting drip, leaving on the surface from which they have retired the moss flattened and blackened as if burnt by fire. With this moss I have rather a curious association, with a description of which it may be worth while to wind up my desultory remarks, as a specimen of what the botanist may have sometimes to experience in his pursuit of Alpine plants. Some years ago, while botanizing with a friend over the Breadalbane mountains, we found ourselves, a little before sunset, on the summit of Ben Lawers, so exhausted with our day's work that we were utterly unable to descend the south side to the inn at the foot, and resolved to bivouac on the hill for the night. The sappers and miners of the Ordnance Survey, having to reside there for several months, had constructed square open enclosures, like sheep-folds, in the crater-like hollow at the top, to shelter them from the northern blasts. In one of these roofless caravansaries we selected a spot on which to spread our couch. Fortunately, there was fuel conveniently at hand in the shape of bleached fragments of tent-pins and lumps of good English coal, proving that our military predecessors had supplied themselves in that ungenial spot with a reasonable share of the comforts of Sandhurst and Addiscombe; and my companion volunteered to kindle a fire, while I went in search of materials for an extemporaneous bed. As heather, which forms the usual spring-mattress of the belated traveller, does not occur on the summits of the higher hills, we were obliged to do without it—much to our regret; for a heather-bed (I speak from experience) in the full beauty of its purple flowers, newly gathered, and skilfully packed close together, in its growing position, is as fragrant and luxurious a couch as any sybarite could desire. I sought a substitute in the woolly-fringe moss, which I found covering the north-west shoulder of the hill in the utmost profusion. It had this disadvantage, however, that, though its upper surface was very dry and soft, it was beneath, owing to its viviparous mode of growth, a mass of wet decomposing peat. My object, therefore, was so to arrange the bed that the dry upper layer would be laid uniformly uppermost; but it was frustrated by the enthusiasm excited by one of the most magnificent sunsets I had ever witnessed. It caused me completely to forget my errand. The western gleams had entered into my soul, and etherealized me above all creature wants. Never shall I forget that sublime spectacle; it brims with beauty even now my soul. Between me and the west, that glowed with unutterable radiance, rose a perfect chaos of wild, dark mountains, touched here and there into reluctant splendour by the slanting sunbeams. The gloomy defiles were filled with a golden haze, revealing in flashing gleams of light the lonely lakes and streams hidden in their bosom; while, far over to the north, a fierce cataract that rushed down a rocky hill-side into a sequestered glen, frozen by the distance into the gentlest of all gentle things, reflected from its snowy waters a perfect tumult of glory. I watched in awe-struck silence the going down of the sun, amid all this pomp, behind the most distant peaks—saw the few fiery clouds that floated over the spot where he disappeared fade into the cold dead colour of autumn leaves, and finally vanish in the mist of even—saw the purple mountains darkening into the Alpine twilight, and twilight glens and streams tremulously glimmering far below, clothed with the strangest lights and shadows by the newly risen summer moon. Then, and not till then, did I recover from my trance of enthusiasm to begin in earnest my preparations for the night's rest. I gathered a sufficient quantity of the moss to prevent our ribs suffering from too close contact with the hard ground; but, unfortunately, it was now too dark to distinguish the wet peaty side from the dry, so that the whole was laid down indiscriminately. Over this heap of moss we spread a plaid, and lying down with our feet to the blazing fire, Indian fashion, we covered ourselves with another plaid, and began earnestly to court the approaches of the balmy god. Alas! all our elaborate preparations proved futile; sleep would not be wooed. The heavy dews began to descend, and soon penetrated our upper covering, while the moisture of the peaty moss, squeezed out by the pressure of our bodies, exuded from below; so that between the two we might as well have been in "the pack" at Ben Rhydding. To add to our discomfort, the fire smouldered and soon went out with an angry hiss, incapable of contending with the universal moisture. It was a night in the middle of July, but there were refrigerators in the form of two huge masses of hardened snow on either side of us; so the temperature of our bedchamber, when our warming-pan grew cold, may be easily conceived. For a long while we tried to amuse ourselves with the romance and novelty of our position, sleeping, as we were, in the highest attic of her Majesty's dominions, on the very top of the dome of Scotland. We gazed at the large liquid stars, which seemed unusually near and bright; not glimmering on the roof of the sky, but suspended far down in the blue concave, like silver lamps. There were the grand old constellations, Cassiopeia, Auriga, Cepheus, each evoking a world of thought, and "painting, as it were, in everlasting colours on the heavens the religion and intellectual life of Greece." Our astronomical musings and the monotonous murmurings of the mountain streams at last lulled our senses into a kind of doze, for sleep it could not be called. How long we lay in this unconscious state we knew not, but we were suddenly startled out of it by the loud whirr and clucking cry of a ptarmigan close at hand, aroused perhaps by a nightmare caused by its last meal of crude whortleberries. All further thoughts of sleep were now out of the question; so, painfully raising ourselves from our recumbent posture, with a cold grueing shiver, rheumatism racking in every joint, we set about rekindling the fire, and preparing our breakfast. In attempting to converse, we found, to our dismay, that our voices were gone. We managed, however, by the help of signs, and a few hoarse croaks, to do all the talking required in our culinary conjurings; and, after thawing ourselves at the fire, and imbibing a quantity of hot coffee, boiled, it may be remarked, in a tin vasculum, we felt ourselves in a condition to descend the hill. A dense fog blotted out the whole of creation from our view, except the narrow spot on which we stood; and, just as we were about to set out, we were astonished to hear, far off through the mist, human voices shouting. While we were trying to account for this startling mystery in such an unlikely spot and hour, we were still more bewildered by suddenly seeing, on the brink of the steep rocks above us, a vague, dark shape, magnified by the fog into portentous dimensions. Here, at last, we thought, is the far-famed spectre of the Brocken, come on a visit to the Scottish mountains. Another, and yet another appeared, with, if possible, more savage mien and gigantic proportions. We knew not what to make of it. Fortunately, our courage was saved at the critical moment by the phantoms vanishing round the rocks to appear before us in a few minutes real botanical flesh and blood, clothed, as usual, with an utter disregard of the æsthetics of dress. The enthusiasm of our new friends for Alpine plants had caused them to anticipate the sun, for it was yet only three o'clock in the morning.

CHAPTER II.

THE INTERMEDIATE OR HEATHER REGION.

THE botanist regards the rapid progress of agriculture in these days with feelings somewhat akin to those which once convulsed the placid bosoms of the Lake poets at the prospect of that "insane substruction," a railway amid the beautiful solitudes of Windermere. He sees, with a sinking of the heart, which no hope of increased gain to the neighbouring gastric region can allay, the wave of cultivation stealthily creeping up the hill-side higher and higher with each yearly tide. The beautiful green knolls around which superstitious eyes used to see the fairies dancing in the midsummer moonlight have been levelled and taken in as part of the surrounding cornfield. The grey Druidical stones which our ancestors reverently spared, and around which the most grasping farmer used to leave a broad margin of natural sward, have been blasted to macadamize a road or build a dyke, in defiance of the curse pronounced against those who should desecrate these old bones of an extinct faith; and the ground on which they stood has been planted with potatoes or turnips. From this universal utilization of the soil the poet and painter have suffered, but not to the same extent as the botanist; for, besides the loss of the beautiful and the picturesque, he has to deplore the gradual diminution of the number of his favourite wild flowers in this country. Meres and lochs in which local aquatic plants luxuriated have been drained, woods have been cut down, and railways and highroads carried through nooks that sheltered the last survivors of an ancient flora. For the extermination of these interesting rarities no quantity of weeds introduced among seed from other countries can compensate.

With these conservative instincts I deeply sympathise; but I rejoice that, while some injury has been done in certain places to special studies, far more land has been left untouched than has been "improved." As ocular demonstration is more convincing than any amount of logical argument, let me ask the botanist who is groaning amid the wheat and turnip fields of the midland counties to accompany me to the top of a hill, say in the highlands of Perthshire. From this superior standing-point let him look around, and he will be at once convinced of the utter groundlessness of his botanical fears. How vast the dominion of Nature! how insignificant the portion that has been reclaimed! For all the evidence of man's occupancy that appears within the boundless horizon, he might imagine himself the solitary tenant of an alien world, monarch of all he surveys. A few spots of pale green hardly seen among the heather; a narrow strip of cultivated valley obscured by the shadow of overhanging mountains; the silver thread of a stream running through a thin fringe of verdure; and, all around, the brown interminable wastes lengthening as he gazes, until their wild billows subside on the blue shore of the distant horizon! This is what he sees, and a more humbling spectacle I cannot imagine. The powerlessness of man's efforts amid the stern forces of Nature could not be more strikingly exhibited. The most rabid opponent of utilitarianism will own that a few scratches, more or less, of the plough, however important to man, are of very little consequence amid these immeasurable deserts.

Nature takes ample care of her own rights. In the rigour of her climate and the ruggedness of her soil she imposes barriers upon the onward march of improvement which cannot be overleaped. It will not pay to cultivate the largest portion of our country. The most powerful artificial manures, and the most skilful "high farming," will not suffice to extract a remunerative produce from our more elevated hills and moorlands. Whatever the pressure of population may be, we must leave these solitudes to their primitive wildness, and give them over in fee-simple to the grouse and Alpine hare. They are the last strongholds into which beleaguered Nature, everywhere else subdued, has withdrawn behind her glacis and battlements of mountain ridges in grim defiance of the advancing conqueror. Nor is it difficult to find reasons for putting up this trespass-notice and restricting man's occupancy of the earth. The lofty mountain ranges have been piled up, and the rugged desolation of the moorlands spread out, because the soul requires some great outlets of this kind to escape from the petty cares and conventionalities of civilized life, and to expand in sublime imaginings towards the infinity of God. While to those who do not feel this craving for something higher and purer than they find in the every-day pursuits of life, and who, like good Bishop Burnet, consider hills and moors unsightly excrescences and deformities upon the face of nature—evidences of the ruinous effects of the Fall—it may be sufficient to say, in justification of this reckless waste of land, that there is a physical as well as an æsthetic necessity for it. There is vicarious sacrifice in the arrangements of inanimate nature, as well as in the laws of human life. There is a beautiful balance by which barrenness is set over against fertility, and life against death. Some spots must be bleak and desolate in order that other spots may be clothed with verdure and beauty. These hills and moors are intended to be not only ornamental, but useful; not only picture-galleries for the poet and painter, but also storehouses of fertility and wealth for the farmer and merchant. Their towering crests and spongy heaths arrest the vapours which float in the higher regions of the atmosphere, collect and filter them in reservoirs in their bosoms, and send them down in copious streams to water the low grounds, and spread over the barren plains the rich alluvium which they bear away in solution from their sides; while the fresh cool breezes, that play around the summits, sweep down with healthful influences into the hot and stagnant air of the confined valleys. In many ways they perform a most important part in the economy of nature, and by their means is preserved the fertility of extensive regions which would otherwise become hopelessly sterile.

To those who are accustomed to the rich beauty of lowland scenery the treeless, desolate aspect of the moorlands may appear harsh and uninviting. They miss there the objects which they are accustomed to see, and around which have gathered the associations of years. There is apparently nothing within the circle of vision to arrest the eye or interest the mind. All seems one dead dull monotony, an interminable dark level, an eye-wearying waste, marked only but not relieved by grey rocks and shallow bogs reflecting an ashen sky. This first unfavourable impression, however, is sure to be dispelled by a more intimate acquaintance. Apart from the charm of contrast which most persons find in circumstances differing widely from those in which their life is usually spent, and the interest which contemplative minds find in all bare, solitary places, there are countless objects of attraction and beauties of hue and form which fill up the seeming void, and make these apparently blank pages of nature most suggestive even to the dullest intellect. The seasons, marching with their slow solemn steps over the moorlands, may leave behind them none of those striking changes which mark their progress in the haunts of man. The elements of the scenery are too simple to be very susceptible to the vicissitudes of the year. But, still, there are some tokens of their presence; and these are all the more interesting that they do not reveal themselves at once to a cold casual gaze, but require reverently to be sought out. Nowhere is the grass so vividly green in early spring-time as along the banks of the moorland stream, or on the shady hill-side, on which the cloud reposes its snowy cheek all day long and weeps away its soul in silent tears. How gorgeous is that miracle of blossoming when Summer with her blazing torch has kindled the dull brown heather, and every twig and spray burst into blushing beauty, and spread wave after wave of rosy bloom over the moors, until the very heavens themselves catch the reflection, and bend enamoured over it with double loveliness! How rich, under the mild blue skies of Autumn, are the russet hues of the withered ferns and mosses that cluster on the braes or creep over the marshes, imparting a mimic sunshine to the scene in the dullest day! How exquisitely pure is the untrodden snow in the hollows which the winds heap into gracefully swelling wreaths and mark with endless curves of beauty! Wander over one of the Perthshire moors from break of morn to close of day, and you will no longer stigmatize it as a monotonous uninteresting waste. From sunrise to sunset the appearance of the landscape is never precisely the same for two successive hours. Like a human face, changing its expression with every thought and feeling, it alters its mood as cloud or sunshine passes over it. Now it is bathed in light, under which every cliff and heather-bush shine out with the utmost distinctness; anon it lies cold and desolate, unutterably forlorn and forsaken when the sky is overcast. At one time it is invested with a transparent atmosphere in which the commonest and meanest objects are idealized as in a picture; at another, great masses of sharply-defined shadows from the stooping clouds lie like pine-forests on the bright hill-sides; or a flood of molten gold, welling over the brim of a thunder-cloud, streams down and irradiates with concentrated glory a single spot, which gleams out from the surrounding gloom like a lovely isle in a stormy ocean. And the sunrises and sunsets—those grand rehearsals of the conflagration of the last day—who can describe them in an amphitheatre so magnificent, a region so peculiarly their own! How inexpressibly sweet is the lingering tremulousness of the gloaming, that quiet ethereal Sabbath-like pause of nature in which the smallest and most distant sounds are heard, not loud and harsh, but with a fairy distinctness exquisitely harmonized with the holiness of the hour! There are no such twilights in England; they belong only to northern latitudes, where the light, if it be colder and feebler, compensates by its longer stay, and its heavenly purity and beauty at the close. And how full of weird, wild mystery is the scene as the evening grows darker; how vast and vague and awful in the uncertain light are the forms of the hills; how ghostly are the shadows! There Night is a visible form, and her solitude is like the presence of a god.

Nor is the moorland altogether dependent for its beauty upon atmospheric effects. It hides within its jealous embrace many a lovely spot on which one comes unexpectedly with all the interest of discovery. There are little dells where a streamlet has lured up from the valley, by the magic of its charms, a cluster of rowan-trees, whose red berries dance like fire in the broken foam of the waterfalls, or a group of tiny, white-armed birches that always seem to be combing their fragrant tresses in the clear mirror of its linns. There are moorland tarns, sullen and motionless as lakes of the dead, lying deep in sunless rifts, where the very ravens build no nests, and where no trace of life or vegetation is seen—associated with many a wild tradition, accidents of straying feet, the suicide of love, guilt, despair. And there are lochs beautiful in themselves, and gathering around them a world of beauty; their shores fringed with the tasselled larch, their shallows tesselated with the broad green leaves and alabaster chalices of the water-lily; and their placid depths mirroring the crimson gleam of the heather hills and the golden clouds overhead.

I have often been struck, when wandering over the moors, with the wonderful harmonies of the various objects. The birds and beasts that inhabit the scene are clothed with fur or plumage of a brown russet hue, to harmonize them with the colour of the heathy wastes, and thus to facilitate their escape from their enemies. Nor is this harmony confined to the form and hue of the living creatures—it is also strikingly displayed in their peculiar cries. All the voices of the moorland are indescribably plaintive—suggestive of melancholy musings and memories. No one can hear them, even on the sunniest day, without a nameless thrill of sadness; and, when multiplied by the echoes through the mist or the storm, they seem like cries of distress or wailings of woe from another world. In them the very spirit of the solitude seems to find expression. None of our familiar songbirds ever wander to the moorland. It is tenanted by a different tribe, and the line of demarcation between them is sharply defined. In the valley and the plain the thrush and the chaffinch fill the air with their music; but, as you climb the mountain-barrier of the horizon, you are greeted on the frontier by the wild cries of the plovers which hover around you in ceaseless gyrations, following your steps far beyond their marshy domains. These are the outposts—the sentinels of the wild—and jealously do they perform their office. No stranger appears in sight, or sets a foot within their territories, without eliciting the warning cry. Well might the Covenanters curse them, for many a grey head, laid low in blood by the persecuting dragoons, would have escaped, securely hidden among the green rushes and peat-bogs, but for their importunate revelation of the secret. Beyond the haunts of this bird stretches a wide illimitable circle of silence, in which only a shrill solitary cry now and then is heard, rippling the stillness like a stone cast into the bosom of a stream, and leaving it, when the wave of sound has subsided, deeper than before. And how absolute is that silence! It seems to breathe—to become tangible. The solitude is like that of mid-ocean—not a human being in sight, not a trace or a recollection of man visible in all the horizon; from break of day to eventide no sound in the air but the sigh of the breeze round the lonely heights, the muffled murmur of some stream flashing through the heather, or the long, lazy lapse of a ripple on the beach of some nameless tarn.

Here, if anywhere, you can be lulled on the lap of a placid antiquity. These grey northern moors are immeasurably old. The gneissic rock that underlies them is one of the oldest in the records of geology—the lowest floor of the most ancient sea, in whose water its particles were first precipitated, to be afterwards indurated by chemical action, or mechanical pressure, into their present compact mass. Here was, probably, the first dry land that appeared above the surface of the ocean. Long before the Alps upreared their snowy peaks from the deep, and while an unbroken sea tossed its billows over the spots where the Andes and Himalayas now tower to heaven, these moors lay stretched out beneath the disconsolate skies, as islands reposing on a shoreless ocean, not clothed, as at present, with brown heather and spongy moss, but presenting an aspect of still drearier desolation. They were all that in the earliest geologic epochs represented the beauty and power of Great Britain—the first instalment of that mighty empire which Britannia gained from the deep. Here, where Nature is all in all and man is nothing, you expect to find permanence. Time seems to have sailed over these moors with folded wing, leaving no more trace of his flight than the passage of the shadow over the dial-stone; and yet, calm and stedfast as the scene may appear, it has passed through many a stormy cataclysm, it has witnessed many a startling transition. On rock and mound the careful observer will find those strange hieroglyphics in which Nature's own hand has chronicled the eventful history of her youth. Here, where the sheep are quietly nibbling the green sward, the sea once broke in foam on the shore; there, on that elevated knoll—if the surface were fully exposed—veins of granite thrust up by some violent internal convulsion might be seen reticulating the gneiss as with a gigantic network, showing the mighty levers employed by Nature in piling up her Cyclopean masonry. Yonder the rocks are smoothed and polished, or else marked with grooves and scratches, telling of glaciers that passed over them, and suggesting to the imagination the picture of that strange era in the past history of our country, when from Snowdon and the Yorkshire moors to Ronaldsay and Cape Wrath eternal winter reigned with sternest rigour, and the Arctic bear hunted the narwhal amid the icebergs and icefloes that drifted past the coasts of Sussex and Hampshire. Yonder granite boulders that strew the hill-side, differing in mineral character from the prevailing formation of the region, and which, according to the Ossian mythology, fell from the leaky creel of a giant Finn striding over the heights one day to take vengeance with this rude but effective ammunition against an offending neighbour, the geologist tells us were transported to this place from a granitic district twenty miles distant on the back of a slow-moving glacier. And the elevated conical mounds, or moraines, which you meet with here and there, are accumulations of mud and gravel, marking in enduring characters the terminations of those vanished ice-streams. Turning from the distant silent ages of the geologist to the early lisping ages of our own race, we find numerous traces of these also chronicled on the moors. The labour of the peasant often discloses, deeply embedded in the moss, large trunks of birch, alder, and fir, masses of foliage, cones and nuts in a perfect state of preservation, the fossils of the peat-bog. These, like the kindred relics of the coal-fields, tell us a tale of luxuriant forests clothing, like dark thunder-clouds, desolate tracts where not a single tree is now to be seen, and scarcely a juniper-bush can grow. Through the underwood of these primæval forests the wild boar roamed, and the shaggy bison bellowed, and the long dismal howl of the wolf made the silence of midnight hideous, ages before the fanfare of the Roman trumpets startled the echoes of the hills. Nor are the traces of man's own presence in those remote times absent from the scene. The sides of some of the hills, which time out of mind have been abandoned irretrievably to the dusky heather, bear evident marks of tillage; but the comparative fertility of these stony spots only proves the wretched state of the agriculture of the Aborigines. Here and there you stumble upon a grey moss-grown obelisk, a cairn, or a cromlech—dim and undated relics, lying, like the fragments of an old world, on the twilight shores of the sea of time. Beside or under these we find the hatchet of stone, the arrow-head of flint, or the quern, over which no history or tradition sheds light. Who owned these rude implements? We cannot tell. Every recollection of the people who used them is swept away. Under the cromlech or the cairn they lay down and took their long, last sleep, without a thought of posterity, or a care as to the conclusions future ages might arrive at regarding the scanty memorials they left behind.

The vegetation of the moorlands is exceedingly varied and interesting. Its character is intermediate between the Arctic and Germanic type, reminding one, in the prevalence of evergreen, thick, glossy-leaved plants, of the flora of Italy, which seems, from the evidence of ancient records, to have undergone a remarkable change in modern times, and now approximates in its general physiognomy to the flora of dry mountain regions. The plant which above all others is characteristic of the moor is, of course, the common heather or ling. It is one of the most social of all plants, covering immense tracts with a uniform dusky robe, and claiming, like an absolute autocrat, exclusive possession of the soil. And yet, though capable of growing in the bleakest spots, and enduring the utmost extremes of temperature, its distribution in altitude and latitude is singularly limited. It ascends only to a certain height on the mountains on which it grows; for, although it covers the summits of most of the hills in England, many of the loftiest Highland hills rise high above it, green with grass, or grey with moss and lichens. Its upper line runs from two to three thousand feet in the counties of Perth, Aberdeen, and Inverness, varying according as it grows on an elevated mountain range or on isolated peaks. On the west coast of Scotland it is very often found on a level with the sea-shore, almost mingling with the dulse and the bladder-wrack. In Norway, strange to say, although the general surface of the country is composed of high and barren plateaux, it is so scarce and local that one may travel hundreds of miles without finding a single specimen. It is replaced in such localities by the bearberry and crowberry, which form immense continuous patches, and look at a distance, especially when withered, in spring or autumn, somewhat like heather. Although abundant on the European side of the Ural mountains, it disappears very suddenly and decidedly on the eastern declivity of the range; and it is entirely absent from the whole of Northern Asia to the shores of the Pacific. Its northern limits seem to be in Iceland, and its southern in the Azores. In Europe it covers large tracts of ground in France, Germany, and Denmark, particularly in the landes of Bordeaux and the moors of Bretagne, Anjou, and Maine; while in Great Britain it exists in every county, with the exception of Berks, Bucks, Northampton, Radnor, Montgomery, Flint, Lincoln, Ayr, Haddington, Linlithgow. The range of the heath tribe is eminently Atlantic, or Western. It is found along a line drawn from the north of Norway along the west coast of Europe and Africa, down to the Cape of Good Hope, in the vicinity of which the family culminates in point of luxuriance of growth, beauty of flowers and foliage, and variety of species, some even attaining the arborescent form. Along this line, which is comparatively narrow, seldom running far from the coast, about four hundred distinct kinds, excluding varieties, are scattered, of which

England and Scotland possess only four, and Ireland no less than six.

On the barren moors of Cornwall a very interesting kind of heather, called the Cornish heath (*Erica vagans*), grows abundantly, distinguished by its crowded bell-shaped flowers. On the north coast of the same county another species occurs, called *Erica ciliaris*, with very large and gaily-coloured flowers, and leaves elegantly fringed with hairs. It is frequent near Truro and Penrhyn, and in one or two places in Dorset. These two Cornish heaths are also found in Ireland; the one on a little island off the coast of Waterford, and the other near Clifton in Galway. In the Emerald Isle, Mackay's heather, which has large glabrous foliage, with an unusual proportion of white under-surface, grows in one or two spots in Connemara. It was discovered the same year on the Sierra del Peral, in Spain. In mountain-bogs in the west of Mayo and Galway the Mediterranean heather is sparingly distributed, sometimes attaining a height of five feet, with numerous upright rigid branches, and flowers in leafy racemes. The Scottish Menziesia (*M. cærulea*), the most abundant kind of heath in Norway, is, as I have already said, almost extinct on Dalnaspidal moor in Perthshire, its only locality in this country. Every visitor in Ireland must be familiar with St. Dabeoc's heath (*Menziesia polli-folia*), which the guides and peasants frequently sell to tourists at exorbitant rates, as a memorial plant. This lovely heather occurs in great profusion on the low granitic hills to the westward of Galway, all the way from the lower end of Lough Corrib. It grows on the heathy moors by the roadsides, and though it is found a considerable way up the mountains, it is there much less abundant, smaller in size, and rarely flowers. The common Bell heather of our Highland moorlands (*Erica cinerea*) produces the finest effect of all our native heaths, growing as it does in great masses in bare places, especially where the burning of the common ling has enriched the soil with its ashes, and removed a formidable competitor in the struggle of existence. It frequently purples a whole hill-side; and nothing finer, as regards effect of colour, can be seen even in the tropics. Mr. Wallace, in his recent work on "The Malay Archipelago," says: "The result of my examinations has convinced me that the bright colours of flowers have a much greater influence on the general aspect of nature in temperate than in tropical climates. During twelve years spent amid the grandest tropical vegetation, I have seen nothing comparable to the effect produced on our landscapes by gorse, broom, heather, wild hyacinths, hawthorn, purple orchises, and buttercups." The cross-leaved heath (*Erica tetralix*) is much less abundant, growing in boggy places among the yellow spikes of the asphodel and the snowy plumes of the cotton-grass. It is more like a hot-house heath, with its rich clustered head of pale rosy blossoms. But growing sparingly, and its colour being more delicate, its effect in the mass, and at a distance, is not equal to its individual beauty close at hand. These two heaths are badges of Highland clans.

That Australia and America have no true heaths is a botanical aphorism. In Australia the tribe is replaced by the *Epacridæ*, which are often as beautiful as any of the Cape heaths. In North America the Scottish Menziesia is more abundant than it is in Scotland, or even in Norway. That continent possesses many plants that are closely allied to the heath tribe. *Hudsonia ericoides*, which covers the white sandy wastes in many parts of New Jersey, is so like the common heath that it is not unfrequently mistaken for it when out of flower. And in the immense forests which clothe every hill and dale of the Laurel, Greenboy, and Alleghany range, rhododendrons, kalmias, azaleas, andromedas, and other plants of the heath alliance, form the chief underwood, and are remarkable for their size and age. It is recorded of the first Highland emigrants to Canada, that they wept because the heather, a few plants of which they had brought with them from their native moors, would not grow in their newly-adopted soil. It is understood, however, that an English surveyor, nearly thirty years ago, found the common ling in the interior of Newfoundland; while in one spot in Massachusetts it occurs very sparingly over about half an acre of boggy ground, in the strange company of andromedas, kalmias, and azaleas peculiar to the country. It was first observed ten years ago, by a Scottish farmer residing in the vicinity, who was no less surprised by its unexpected appearance than delighted to set his foot once more on his native heath. None of the plants seemed to be older than six years, and may, therefore, have been introduced by some one who found relief from home sickness in forming this simple floral link between the new and the old country.

There are many beautiful little shrubs growing on the moorland along with the heather which are found nowhere else. The crowberry spreads over rocky places in large tufted masses, producing early in summer a liberal supply of black juicy berries, which form the principal food of the grouse and other moorland birds. The dry barren knolls, where the wind blows keenest and the scent of water is never felt, are profusely covered with the trailing stems and glossy leaves of the bearberry. The flower is even more beautiful than that of either the cross or fine-leaved heather—a little waxen bell, with the faintest blush on its snowy cheeks; and the fruit is no less lovely, clusters of mealy beads of the richest crimson gleaming out in beautiful contrast from the dark green leaves. A species called the black bearberry is found on dry barren grounds on many of the Highland mountains. The flowers are of a pale rose colour, and the berries of a rich lustrous black. On Ben Nevis, near the lake, on Hoy Hill, Orkney, and especially on the mountains of Sutherland and Caithness, this rare species occurs, and forms an attractive feature in the Alpine landscape towards the end of autumn, when its leaves assume a brilliant flame colour. The famous strawberry-tree, or *Arbutus*, so conspicuous in the beautiful scenery of Killarney, and supposed by some to have been introduced from Spain by the monks of Mucross Abbey, is an arborescent form, an aristocratic relative of this lowly Highland family. On the moist hillsides the mountain rasp or cloudberry, the badge of the clan Macfarlane, grows in great abundance; and its rich orange fruit, under the name of *eiracan* or *noops*, furnishes a grateful refreshment to the shepherd on a hot autumn day. One of the most beautiful plants of the moorland is the Marsh Andromeda. It is found but sparingly in a few places in the North of England and in the Lowlands of Scotland, and in Queen's County and Kerry, Ireland; but where it is found, it is a prize worth going for to get. It is a small evergreen shrub, with oval ruby-coloured flowers concealed among the terminal leaves. In Norway it is very abundant on the moors in company with the Menziesia. I gathered it in great profusion by the roadsides when passing through Romsdal, between Nystven and Ormen; its rose-coloured flowers fringing the ditches and peeping out from among the boulders. The beauty of its flowers when contrasted with the dreariness of its habitats, supposed to be haunted by supernatural beings, led to its receiving the classical name of the beautiful virgin who was chained to the rock and exposed to the attack of the sea-monster. Another beautiful plant common on the Highland moorlands is the *Pyrola* or winter-green, which loves to grow in upland pine-woods, or under the lee of some dense heather bush, perfuming the air when it occurs in any quantity with its delicate scent, strongly suggestive of the lily of the valley. In similar situations the bilberry also luxuriates. Abundant everywhere on the exposed sides of the hills, it flowers and fruits only in the shelter of the woods or on the shady banks of subalpine streams. Its berries are exceedingly agreeable to the taste, and are largely used in the form of preserves in the Highlands. Blaeberry hunting in July is a favourite pastime among the children; and for days afterwards the persistent stains of the spoil crimson cheeks, lips, and dress. The bog whortleberry is more sparingly distributed, though it is frequent enough on most of the Highland mountains, ascending almost to their summits. The corolla is of a pale rosy colour, and the berry black and juicy, but inferior in flavour to the bilberry. The cowberry (*Vaccinium Vitis Idæa*) ornaments some parts of the Highland mountains, woods, and heaths with its straggling shrubby growth and box-like leaves. It seldom flowers or fruits in this country; but in Norway it bursts into blossom everywhere, and is loaded with pale flesh-coloured flowers, lighting up the dark pine-woods with its beauty. Next to the bilberry, the cranberry is the most interesting and useful of the Vacciniums. It loves moist situations, and therefore occurs in peat-bogs, with its root immersed in the great spongy cushions of the bog-moss, and its evergreen wiry leaves trailing over them. The flowers are of a lovely rose colour, with a deeply divided corolla and segments bent back in a very singular manner. In this country it is very local and scarce; but in Norway it grows in great profusion on almost every hill; and nothing can equal the luxuriousness of its growth and fruiting in the marshes and steppes in the north of Russia, from which the vast quantities used by our confectioners for tarts are annually imported. The juniper forms miniature pine-groves in sheltered places, and yields its berries liberally to give a piquant gin flavour to the old wife's surreptitious bottle of whisky; while the sweet gale or Dutch myrtle perfumes with its strong resinous fragrance the foot that brushes through its beds in the marshes, and gives a similar spice of the hills to the Sunday clothes of the Highland belle, as they are carefully folded with a sprig between each in the "muckle kist." Beneath the shelter of these tiny fruit-trees of the heath there is a dense underwood of minute existences, curious antique forms of vegetable life, performing silently, and all unknown and unnoticed, their allotted tasks in the great household of Nature. The little cup-lichen reddens by thousands every dry hillock; the reindeer-moss whitens the marshes with its coral-like tufts; the long wreaths of the club-moss creep in and out among the heather roots, like lithe green serpents, sewed to the ground by delicate threads, yet sending up here and there from their hiding-places white two-pronged spikes to catch the sunbeams; the sphagnum-moss lines the bogs with its great pads of brilliant crimson or green; and the white fork-moss covers the wet tussocks with its pale cushions, into which the foot sinks up to the ankle; and thus you wander on, observing and gathering each new and strange production, until you are lost in admiration of the wealth of beauty and interest scattered in the waste without any human eye to behold it.

Nor is the moorland altogether destitute of human interest. Far up in some lonely corrie may be seen the ruins of rude sheilings surrounded by soft patches of verdure, on which the heather has not intruded for centuries. To these Highland chalets the wives and daughters of the crofters used to come up from the valley every summer with their cattle and dairy utensils, and spend three or four months in making cheese and butter for the market, or for home consumption during the winter, as is the custom still in some secluded districts of Norway and the Swiss Alps. The Gaelic songs are full of beautiful allusions to the incidents of this primitive pastoral life; and many fresh and interesting materials for poetry or fiction might be gleaned from this source by those who have exhausted every other field. Farther down the hill, though still among the moorlands, there are other ruins of cottages and farmsteads, the effects of those extensive "clearings" which took place forty or fifty years ago in the great Highland properties. Scores of such "larichken," as they are called, with the rank nettle growing round the hearthstone, and surrounded by traces of cultivation, may be seen in places where sheep and deer now feed undisturbed by the presence of man. The wisdom and justice of depopulating these upland valleys have been often questioned. It was, at the least, a terrible remedy for a terrible disease; and we ought, perhaps, as a nation, to be thankful that upon the whole it has been productive of unlooked-for beneficial results. The situation of these ruins is often exceedingly picturesque; perched under the lee of a grey crag, with a little streamlet murmuring past through the greensward, like the voice of memory informing the solitude, and a single fir-tree bending its gnarled branches over the roofless walls, its scaly trunk gleaming red against the sunset, enhancing, instead of relieving, the desolation of the scene. I have spent many happy days in these simple homes, the abodes of honest worth and rough but genuine hospitality, on which I look back through the haze of years with a pleasing regret. Well do I remember your humble hut, Donald Macrae, afar amid the wild moors of Bohespick, with its thatched roof and unmortared walls, green and golden with Nature's lavish adorning of moss and lichen. Your little patch of garden was overgrown with weeds which congregated there from all quarters, as if glad of a shelter from the inhospitable wild, and so rudely fenced in from the heather that the rabbits found easy admission to your peas, and the red deer often came down hunger-driven from the snow-clad heights, and devoured in a few seconds your scanty stock of winter kail; but in no garden of lord or commoner were the red hairy gooseberries so sweet, and Mount Hybla itself could not boast of more luscious honey than the liquid amber gathered from the heather-bells by the three beehives in the sunny corner. I can testify to the noisy welcome of your collies when I used to appear in sight, and to the shyness of your four chubby pledges of affection, as they cautiously peered out at me from behind the safe shelter of the maternal wing, mute and irresponsive to the kindest familiarities, and to the most tempting offers of "sweeties." The vision of your hospitable board rises up before my mental eye, loaded with a pile of crisp oat-cakes; a jug of foaming cream, with that rich nutty flavour peculiar to the produce of cows fed on old pastures uncontaminated by villanous artificial manures; cameos of golden butter, with the national symbol in beautiful relief; a great hard cheese of ewe's-milk; and last, not least, a bottle of native mountain-dew undesecrated by water or gauger's grace. I see dimly, through the peat-reek of your ingle, your own manly face and buirdly figure clad in tartan coat and kilt spun by your aged mother from the fleece of your own sheep, with a collie at your feet, and your youngest hope dandling on your knee, and your comely wife, with mealy cheeks and arms bare to the shoulders, baking the household cakes, as perfect a picture of a Dutch Venus as ever emanated from the pencil of Rubens or Houdekoetter! May the blessing of Him that dwelt in the bush rest upon you and yours in that distant Australian valley, which, true to the instinct of home, you have pathetically named after your native spot!

It is well that there are still many homes of this kind, inhabited by an equally hospitable race, to be found by the stranger when weary and belated in his wanderings amid the Highland moorlands. I know nothing more enjoyable than a week's sojourn in one of these places. The infatuation which drives so many people every season to dissipate their time amid the frivolities of some pert fashionable village or watering-place, on pretence of going to the country, is utterly incomprehensible to me. I would advise every sensible person who wishes a fresh supply of good temper as well as of good health, to avoid carefully, as he would the plague, every one of those spas and villages "within easy reach by coach or railway," and boldly take up his abode in some lonely farmhouse or shepherd's sheiling on the Highland moors. Here, with an utter change of scene, you breathe an air pure and fresh from Nature's own goblet. Ozone, that purifying principle in the atmosphere which is antagonistic to all fevers and miasma, increases with the height; and here it abounds, filling all the atmosphere with its healthful influences. There is a tonic in every draught of it for every species of dyspepsia, for every form of enervation and lassitude that results from a pampered stomach or an overwrought brain. There is balm in every breeze, expanding the spirit and lifting it buoyantly up from under the burden of care and anxiety, until it embraces like a rainbow all nature within its radiant arch, and old cares and sorrows become dim as dreams. You feel as if, besides all the gases needful for respiration, there were present " some ethereal nectarine element baffling the analysis of the chemist," yet revealing its presence in the thrill of conscious exuberant life which it excites in your frame. Here, not far from the centres of civilization, within reach, and yet remote, you may realize the benighted state of our ancestors; feel what it is to exist without letters, newspapers, visitors, calls of ceremony, or any of the thousand and one appliances of modern life, and yet at any time be able to survey from some elevated point a region within whose magic ring all these things are enjoyed. Here is the highest soul of monastic retirement—all its romance, with none of its restraint. You stand apart from the world in an eddy of life, a quiet sheltered bay cut off from the ocean, whose rough stormy waves rave and foam without, with no society save that of the taciturn farmer and his family, the black-faced sheep and the dumb mountains. You will have to put up with some inconveniences, no doubt. You may feel, when forcing your body into the wall-press which stands for your bed in the *ben*-room, as if you were rehearsing, like Charles V.—with the disadvantage of being alive, and no mourners—the ceremony of your own coffining. The friction of the native sheets and blankets against your delicate skin may remind you forcibly of the shampooing which nearly flayed you in a Turkish bath. You will, perhaps, have to wash yourself in the neighbouring burn, in absence of all toilette apparatus. Your diet will be largely a milk one, reducing you to the condition of a Cretan; and your teeth, lately under the care of Messrs. Molar and Co., may have hard work with the granitic cakes and fibrous mutton. But all these disadvantages will enhance, by way of contrast, your enjoyment of the place. They will be incidents to think of pleasantly afterwards amid the luxuries of your club, or during that pleasant half-hour of retrospection before you fall asleep amid the downy billows of civilization's four-poster. And, depend upon it, there will be a great deal of insensible education going on in your converse with your own soul in the solitude of the hills, and a stock of softening influences accumulating, which will make the toilsome dreary days of winter brighter, and prepare you the better for that "bourne from whence no traveller returns."

One of the most frequent incidents of the moorland, about the beginning of June, is peat-making, the most picturesque of Highland outdoor occupations. In those basin-shaped hollows which give the scenery an undulating aspect there are large deposits of peat, formed by the decay of numberless generations of those plants which delight in cool climates and moist soils. The history of this accumulation of carbonaceous matter is exceedingly interesting to the geologist. It furnishes a plausible solution of the difficulties involved in the question of the formation of coal; it provides data by which recent geological changes may be determined with some degree of accuracy; and frequently, owing to its antiseptic qualities, it becomes an archæological cabinet, preserving the relics of former generations. In none of these aspects, however, are the peat-bogs of the Highland moors so interesting as in their connexion with the habits and customs of the peasantry. It is no easy task to thread one's way among the bogs and marshes where the peat is found, the danger being somewhat imminent of falling plump over the yielding edge into some open pool of inky water, or sinking up to the waist in some treacherous spot veiled over with a deceitful covering of the greenest moss. In the outskirts of this wilderness of bogs the peat-makers are hard at work. One man, with a peculiarly shaped spade, cuts the peats from the wall of turf before him and throws them up to the edge of the bog, where a woman dexterously receives and places them on a wheelbarrow, another woman rolling away the load and spreading it out carefully on some elevated hillock, exposed to the sunshine, in order to dry and harden. And thus the process goes on from sunrise to sunset, with an hour's rest for each meal. Though looked forward to, especially by the younger labourers, with much pleasure, as a delightful contrast to the monotony of their ordinary work about the farm, and as affording peculiar facilities for carrying on the mysteries of rustic courtship, peat-making is most fatiguing work; and when, as is often the case, they have to walk a distance of five or six miles to and from the spot, and to carry on their labours under the scorching glare of the sun, exposed without shelter to torrents of rain or piercing winds, it must be confessed that they pay dearly for the materials which in the long cheerless winter of the North afford them both fire and light. In remote inaccessible districts, where wood is scarce and coal almost unknown on account of its enormous price, averaging from 30*s*. to 4*l*. a ton, peat is the sole fuel used by the inhabitants. The whole of a peat-bog, covering in many places an area of several acres, and occupying what was once evidently the bed of a lake, is parcelled out into several portions, which are generally annexed by the proprieter to the holdings of the tenants on his estate who are nearest to the spot. These parcels of peat-bog are usually given free of rent; and the whole expense connected with peats is thus only the labour involved in their manufacture and carriage. So rough are the roads, however, and so long the distances to which they have in most cases to be carried, that peat is not so cheap and economical a fuel as might be supposed. The selling price is usually three shillings a cart, and six carts are understood to last as long as a ton of coal. Peat-making is not nearly so common in the Highlands as it used to be. The facilities of carriage to almost every part of the country by sea and land are now numerous, and coal in consequence is so reduced in price, as to be more within reach of the poorer classes; while the use of that fuel saves time and labour which can be more profitably employed.

Another spectacle peculiar to the moors is the burning of the heather. This practice is not confined to any particular locality, but is followed all over the Highlands. It commences in spring, when the snows have completely disappeared, and the weather is dry and fine, and is carried on at irregular intervals throughout the whole summer. Its object is, by clearing the ground of the heather, under whose shade no other vegetation can grow, to produce pasturage for the sheep. In spots that have been thus cleared the grass grows luxuriantly, and forms a thick close carpet of green verdure, of which the mountain sheep are particularly fond. The stumps of the heather are usually left in the ground, for the fire consumes only the foliage and the smaller twigs; and these skeletons, closely matted together, bleached and sharpened by the elements, frequently crossing one's path, are very disagreeable to walk on, unless the feet are protected by very thick boots. The contrasts of shape and colour formed by these clearings in the aboriginal heather are very curious, and strikingly diversify the monotony of the landscape—here a uniform brown sea of heather; there long stripes of grey colouring running in and out and crossing in all directions, like promontories and capes; and yonder bright green isles of verdure smiling amid the surrounding desolation. The shepherds, unless under the immediate surveillance of a gamekeeper, are often reckless in setting fire to a hill-side, not caring how far the flames may extend, allowing them to burn for days and even weeks, until a friendly deluge of rain extinguishes them. Valuable tracts of grouse moor are thus often ruined beyond repair, and the destructive effects not unfrequently extend to upland woods and cornfields, presenting, on almost an equal scale, a picture of the famous prairie fires of America. Hares and deer are seen careering before the flames; grouse are whirring past blinded and scorched, and lizards and snakes are running hither and thither in an agony of terror; volumes of dense smoke darken the air, and the dull red embers light up the darkness of the night and reflect a volcanic glare upon the surrounding hills. It is one of the grandest sights of the kind to be seen in the Highlands.

These rough, hasty sketches among the heather would be manifestly incomplete without a notice, however brief, of grouse-shooting. Being no sportsman, I despair of giving an adequate conception of the sport to the uninitiated. It is only those who have taken part in it who can understand the importance which it has attained in the world of fashion, and the enthusiasm with which the most phlegmatic English millionaires and members of Parliament enter into it. We have all, from the highest to the lowest, a strong spice of the savage in our nature; and a longing at times comes over us to break loose from the restraints of civilization and revel in the wild freedom of our barbarian ancestors. The grouse-shooting fever may be one of the periodical ebullitions of the original temperament. But, after all, there is really very much to enjoy connected with the sport. The very change from the Babel of noises in the metropolis to the deep hush of Nature's great solitudes has a soothing charm; while the return to simple hardy life is a gratification which is felt all the more keenly, the more that ordinary life is artificial and refined. Then the associations of the sport—the fresh exhilarating air of the hills, laden with the all-pervading perfume of the heather bells; the magnificent prospect of hill and valley stretching around; the blue serenity of the autumnal sky; the carpet of flowering heather glowing for miles on every side, and so elastic to the tread; the vastness and profundity of the solitude; as well as the strange and unfamiliar sights and sounds of the scene—all these appeal to that poetical spiritual faculty which is latent even in the most prosaic statistician of St. Stephen's. Add to these the exciting nature of the sport itself—the feelings of emulation it excites among rival sportsmen; the vigilance and wildness of the birds, requiring the utmost caution and skill in approaching them; the thrill of expectation as the well-trained dogs suddenly stop and point with uplifted paw and anxious look to the spot where a covey is nestled; the sudden startling whirr of the birds ascending at your approach; the satisfaction of bringing down, with well-aimed double fire, the plumpest of them; the rustic luncheon beside the spring; and the return, amid the splendour of the setting sun, with well-filled bag, to be greeted half-way from the snug shooting-lodge, with the warm praises of rosy lips and the fond looks of loving eyes. Nay, even the disappointments to be met with—the long wearisome walks over bog and heather, searching in vain for game; the false pointing of dogs, deceived by the scent left behind in places where game were a while before, but are not now; and the most vexatious thing of all, the defying insolence in the kok-kok-kok of the male bird as he flies off unhurt from your fire at the head of his family—all these are so many elements of the romantic, which throw a halo of the deepest interest around the sport, and make the twelfth of August to be more eagerly anticipated by the weary Londoner than any other day in the calendar. Grouse-shooting has been of incalculable benefit to the Highlands. Thousands of pounds are thus annually spent in the poorest districts; communication is opened up with the most isolated spots; employment is furnished to carriers and gillies, who might otherwise have either to starve or emigrate; and proprietors receive something like a second rent from parts of their estates which were formerly valueless. The preservation of the game is thus of the utmost importance, not unworthy of being considered a national question. Even apart from such selfish considerations, it would be a great pity if this interesting bird should become extinct in the only quarter of the globe where it is found. As it is, every one will be sorry to learn that it is becoming scarcer and wilder every year, disappearing rapidly from localities where it used to be abundant, and now principally confined to the Perthshire and Inverness-shire moors. The only ground of complaint any one can have against the sport is, that it has a tendency to foster that spirit of exclusiveness which characterises many of the great landed proprietors, and induces them to shut up some of the wildest scenery in Scotland from the foot of tourist and savant. The depopulation of many Highland districts through this game mania might be overlooked, owing to the many ulterior advantages that have resulted therefrom, both to those who remain and those who have emigrated. But there is neither advantage nor courtesy in such a strict and extensive application of the law of trespass. The reason commonly alleged for it is a mere pretence. Not one of the true lovers of nature—

and it is only such who would care to penetrate out of the beaten tracks into these spots—but would be as careful of the rights and possessions of the proprietor as though they were his own; and it is difficult to see how the presence at long and rare intervals of a solitary pedestrian in such immeasurable solitudes can have the effect of scaring game. The very worst thing he could do would be merely to send them scudding away from one heather hillock to another; and in all likelihood the human biped would be the more scared of the two by this movement. It requires pretty stout nerves, and somewhat unusual presence of mind, to hear with unruffled composure the sudden and unexpected whirr of a heathcock; while the vision of a herd of wild deer with lowered antlers, in autumn, is sufficient to make the boldest turn tail. Let proprietors enjoy their game rights to the full, but it is unworthy of the liberality of the age to debar the "unlanded" from the enjoyment of universal nature, which to many is as much a necessity as their daily bread, and more than counterbalances the want of property. Full liberty, without any hampering restraint whatever, to wander among the heather, and gather the materials of their study where Nature scatters them with so lavish a hand, should be accorded to the artist and the man of science, whose pursuits do not interfere with the gains or enjoyments of others, and to whom we are indebted for some of the most refined and elevated pleasures of life.

CHAPTER III.

A GARDEN WALL IN A HIGHLAND GLEN.

ALPHONSE KARR, in his charming little work entitled "A Tour round my Garden," shows how much pleasure and instruction may be found by careful eyes and thoughtful minds within the very narrow limits of an ordinary garden, to compensate the sedentary for being deprived of the enjoyments of travel. I have often thought that, if the garden wall, which he has strangely overlooked, were properly described, with all the objects and associations connected with it, the Frenchman's tour would have been made still more interesting. Though one of the most familiar and commonplace objects upon which the eye can rest, it has often suggested to myself many a pleasing and profitable train of thought in dull moods of mind, when least disposed for inquiry or reflection. To those who cannot climb the mountain summit, or wander over the moorland, a few words describing the points of attraction which it possesses may not be out of place at a time when the worker becomes the observer, and serious pursuits are laid aside for a while to enjoy the *dolce far niente* of the country.

Still small voices that were drowned by the bustle of life have now a chance of being heard amid the universal silence; and humble sights of nature—overlooked amid engrossing scenes of human interest—are now appreciated with all the zest of a holiday.

There is a structure before my eye at this moment which is my *beau idéal* of a garden wall. It stands on the brink of a little stream that clothes every mossy stone in its bed with sparkling folds of liquid drapery, and makes its refreshing murmur heard all day long in the garden, animating it as if with the voice of a friend. The space of grassy sward outside between it and the water—green as an emerald—is jewelled with constellations of primroses, anemones, and globe-flowers, as fair in their own order and season as the cultivated flowers which make the borders within gay as the robe of an Indian prince. Three fairy birch-trees bend over it with their white stems glistening like marble columns in the sunlight, and their small scented leaves whispering some sinless secret to the breeze, or, when the wind is hushed, stealing coy glances at the wavering reflection of their beauty in the stream. It is built of rough stones loosely laid above each other without mortar or cement, and coped on the top with pieces of verdant turf taken from the neighbouring common; and would perhaps be considered very unsightly in the suburbs of a city when contrasted with the trim elegant walls surrounding villa gardens. In this situation, however, it is exceedingly appropriate, and harmonizes with the character of the scenery much better than if its stones were chiselled with nicest care, and laid together with all the skill of the architect. The eye of a painter would delight in its picturesqueness, and the accessories by which it is surrounded; for while offering an insuperable obstacle outside to little eager hands, covetous of forbidden fruit, ripe and especially unripe, it is yet sufficiently low inside to permit an unobstructed view of the scenery in front, allowing the eye to wander dreamily over the landscape as it billows away in light and shade —from the green cornfields up to the pine-woods that hang like thunder-clouds on the lower heights —and thence to the brown heather moorlands, and on to the blue hills that melt away in sympathy and peace on the distant horizon. The garden which it surrounds—"the decorated border-land between man's home and Nature's measureless domains"—is very pleasant. Bright with simple old-fashioned flowers, and nestled amid verdure of blossoming tree and evergreen shrub, it looks like a little Eden of peace, sacred to meditation and love, which the noises of the great world reach only in soft and subdued echoes. Alas! the beautifully embroidered robe of nature too frequently reveals the suggestive outlines of some dead joy, though at the same time it mercifully softens over and conceals its ghastly details. There is a sepulchre in this garden too; and, though the wall has been high enough to bound the desires and fancies of simple contented hearts that never sought to mingle in gayer scenes, it has not been sufficiently high to exclude that dark mist of sorrow in which the light of life goes out, and the warmth of the heart gets chill. That wall is dear to me on account of its strangely sweet memories of mingled joy and sadness. Eyes have gazed upon it as a part of their daily vision, that are now closed to all earthly beauty; voices beside it have sounded merrily at the sweetest surprise of the year, when the snowdrop first peered above the sod like the ghost of the perished flowers—voices that suddenly dropped off into silence when our life-song was loudest and sweetest; tender and true hearts under the caresses of its overshadowing birch-trees have known "of earthly bliss the all—the joy of loving and being beloved." Little fingers have often been busy among the flower-beds which it sheltered, leaving touching traces of their work in buds beheaded left lying artlessly beside the parent cluster —joys plucked too soon, and fugitive as they were pleasing. Fresh marks of little teeth have often been found deep sunk in a dozen rosy apples growing temptingly within reach on the lowest bough—a trace of "original sin," natural to every juvenile descendant of Eve, and easy to forgive when, as in these instances, linked with so much innocence; it seemed so childlike to take a bite out of several ripening apples instead of plucking and finishing one. But, apart from such human associations, I have studied the wall often for its own sake; and to me it has all the interest of a volume. Covered over with its bright frescoes of parti-coloured lichens and mosses, and crowned with its green turf, sprinkled with grass-blossoms and gay autumn flowers, it reminds me of the rich binding of an old book on which the artist has bestowed especial care; or rather, it stands in relation to the garden like the quaintly illuminated initial of a monkish chronicle, telling in its gay pictures and elaborate tracery the various incidents of the chapter.

A rough stone wall in any situation is an object of interest to a thoughtful mind. The different shapes of the stones, their varied mineral character, the diversity of tints, flexures, and lines which occur in them, are all suggestive of inquiry and reflection. Sermons may thus be found in stones more profitable, perhaps, than many printed or spoken ones, which he who runs may read. The smallest appearances link themselves with the grandest phenomena; a minute speck supplies a text around which may cluster many a striking thought; and by means of a hint derived from a mere hue or line in a little stone—almost inappreciable to the general eye—may be reconstructed seas and continents that passed away thousands of ages ago— visions of landscape scenery to which the present aspect of the globe presents no parallel. This flexure of the stone tells me of violent volcanic eruptions, by which the soft, newly-deposited stratum—the muddy precipitate of ocean waters—

heaved and undulated like corn in the breeze; that lamination, of which the dark lines regularly alternate with the grey, speaks eloquently of gentle waves rippling musically over sandy shores; and the irregular protuberances, which I see here and there over the stone, are the casts of hollows or cracks produced in ancient tide-beaches by shrinkage—similar appearances being often seen under our feet, as we walk over the pavement of almost any of our towns. Yonder smooth and striated surface of granite is the Runic writing of the northern Frost-king, transporting me back in fancy to that wonderful age of ice when glaciers slid over mountain rocks, and flowed through lowland valleys, where corn now grows, and the snow seldom falls. And if there be a block of sandstone, it may chance to exhibit not only ripple-marks of ancient seas, but also footprints of unknown birds and strange tortoises that sought their food along the water's edge; and sometimes memorials of former things more accidental and shadowy than even these—such as fossil raindrops, little circular and oval hollows, with their casts—supposed to be impressions produced by rain and hail, and indicating by their varying appearances the character of the shower, and the direction of the wind that prevailed when it was falling. Every one has heard of the crazy Greek who went about exhibiting a brick as a specimen of the building which he wished to sell; but in the structure of each geological system every stone is significant of the whole.

Each fragment, however minute, is a record of the terrestrial changes that occurred when it was formed; ingrained in every hue and line is the story of the physical conditions under which it was produced. The Ten Commandments were not more clearly engraved on the two tables of stone than the laws of nature that operated in its formation are impressed upon the smallest pebble by the wayside. Its materials furnish an unmistakeable clue to its origin, and its shape unfolds its subsequent history. God has impressed the marks of the revolutions of the earth not merely upon large tracts of country and enormous strata of rock and mountain range—difficult of access and inconvenient for study—but even upon the smallest stone, so that the annals of creation are multiplied by myriads of copies, and can never be lost. Man cannot urge the excuse that he has no means of knowing the doings of the Lord in the past silent ages of the earth, that His path in the deep and His footsteps in the great waters are hopelessly unknown. Go where he may, look where he pleases, he will see the medals of creation—the signet marks of the Almighty—stamped indelibly and unmistakeably upon the smallest fragments of the dumb, dead earth; so that if he should ungratefully hold his peace, and withhold the due tribute of praise to the Creator, "the very stones would immediately cry out." Anatomists of scenery, who look beneath the surface to the skeleton of the earth, tell us that the features of mountains and valleys are dependent upon the geological character of their materials; and, therefore, those who are skilful in the art can tell from the outlines of the landscape the nature of the underlying rocks, although no part of them crop above ground. A passing glance at the wayside walls will reveal the prominent geology of any district, just as the shape of a single leaf and the arrangement of veins on its surface suggest the appearance of the whole tree from which it has fallen, or as a fragment of a tooth or a bone can call up the picture of the whole animal of whom it formed a part. In Aberdeenshire, the walls are built principally of granite, grey and red; in Perthshire, of gneiss and schist; in Mid-Lothian and Lanarkshire, of sandstone; and in the southern Scottish counties generally, of trap and porphyry. Sometimes they are composed of transported materials, not native to the district; and the history of these opens up a field of delightful speculation. But there are no walls so interesting as those which occur in the mountain districts of Derbyshire, and in some parts of Lancashire. In almost every stone are embedded fossil shells, and those beautiful jointed corals called encrinites, which look like petrified lilies, and have no living representatives in the ocean at the present day. Even the most homogeneous blocks are found on close inspection to be composed entirely of mineralized skeletons, and to form the graves of whole hecatombs of shells and corallines long ago extinct. Strange to think that our limestone rocks are formed of the calcareous matter secreted by living creatures from the waters of the sea, and their own shelly coverings when dead, just as our coal-beds are the carbonized remains of former green, luxuriant forests. Thus, while walking along the highway in almost any locality, the most hasty examination of the wall on either side furnishes the student of nature with abundant subjects for reflection; and those lofty dykes, built by the farmer to keep in his cattle, or by the jealous proprietor to secure the privacy of his domain, while they forbid all views of the surrounding country, amply compensate for the restriction they impose by the truths engraven on their seemingly blank but really eloquent pages—like the tree which in winter permits us to see the glory of the sunset and the purple mountains of the west through its lattice-work of boughs, but in summer confines our vision by the satisfying beauty of its full foliage and blossoms.

The mist of familiarity obscures, if not altogether hides, the intrinsic wonder that there is about many of our commonest things. The existence of stones is an accepted fact, suggestive of no thought or feeling—unless, indeed, we stumble against one; we look upon them as things of course, as natural in their way as the rocks, streams, and woods around—as a necessary and inevitable part of the order of creation; and yet they are in reality well calculated to excite curiosity. Sterling, in his " Thoughts and Images," beautifully says: " Life of any kind is a confounding mystery; nay, that which we commonly do not call life—the principle of existence in a stone or a drop of water—is an inscrutable wonder. That in the infinity of time and space anything should be, should have a distinct existence, should be more than nothing! The thought of an immense abysmal nothing is awful, only less so than that of All and God; and thus a grain of sand, being a fact, a reality, rises before us into something prodigious and immeasurable—a fact that opposes and counterbalances the immensity of non-existence." But this wonder and mystery stones share in common with all material things; their own origin is a special source of interest. Many individuals, if they think at all about the subject, dismiss it with the easy reflection that stones were created at first precisely in the form in which they are now found. It may, however, be laid down as a geological axiom, that no stones were originally created. The irregular aggregations of hardened matter so called formed part at first of regular strata and beds of rock, and were broken loose from these by volcanic eruptions, by the effects of storms or floods, by frost and ice, or by the slow corroding tooth of time. By these natural agencies the hard superficial crust of the earth has been broken up into fragments of various sizes, carried away by streams, glaciers, and landslips—modified in their shapes by friction against one another, and at last, after many changes and revolutions, deposited in the places where they are found. We owe the largest proportion of the stones scattered over the surface of the earth to glacial action—one of the most recent and remarkable revolutions in the annals of geology. Man is thus provided with materials for building purposes conveniently to his hand, without the necessity of blasting the rock, or digging into the earth; and it is a striking thought, that the very same great laws by which the disposition of land and sea has been effected, and the great features of the earth modified, have conduced in their ultimate results to the homeliest human uses. The materials which the poorest cotter builds into the rudest crowfoot dyke around his kail-yard or potato-field, have been produced by causes that affected whole continents and oceans. The meanest and mightiest things are thus intimately associated and correlated; just as the forces that control the movements of the stars are locked up in the smallest pebble—keeping its particles together, a miniature world. Stones are sometimes out of place, as when they occur in a field or garden; but they form a feature in the æsthetic aspect of scenery which could not well be wanted. What a picturesque appearance do the huge rough boulders strewn over its surface impart to the green hill-side! especially if, as is often the case, their sides are painted and cushioned with that strange cryptogamic vegetation which one sees nowhere else, and a daring rowan-tree plants itself in their crevices and waves its green and crimson flag of victory over soil and circumstances.

There are few things more beautiful than the pebbly beach of a mountain lake; and some of the finest subjects for a picture may be found by the painter along the rough, rocky course of a mountain stream, where the stones form numerous snowy waterfalls, and the spray nourishes hosts of luxuriant mosses and wild flowers. Although dumb, and destitute of sonorous properties, how large a share of the sweet minstrelsy of nature is contributed by them. They are the strings in the harp of the stream, from which the snowy fingers of the water-nymph draw out ever-varying melody—a ceaseless melody, heard when all other sounds are still. By their opposition to the current they create life and music amid stillness and monotony, change the river from a dull flat canal into a thing of wild grandeur and animation, and redeem the barren waste from utter silence and death. Commonest of all common things, it is strange to think that there are parts of this rocky material earth of ours where stones are as rare as diamonds, and the smallest pebble is a geological curiosity. The natives of some of the coral islands of the Pacific procure stones for their tools—this being the only purpose for which they use them—solely from the roots of trees that have been carried away, with their load of earth and stones adhering to them, by the waves from the nearest mainland, and grounded upon their shores. So highly are these stray waifs of the ocean valued that a tax is laid upon them, which adds considerably to the revenue of the chiefs. This reminds us of the preciousness of stones during what is called the stone age of our own country—whose date is so apocryphal—when flint and granite were the sole materials employed for making the various implements of war and of household use, and these rude implements were buried with the dead in the stone cist under the huge cromlech or grey cairn. Those relics dug up in the times of our forefathers, before the attention of antiquaries or geologists was directed to the subject, were accounted as holy stones, supposed to have formed part of the cabalistic appendages of the necromancer of bygone ages; and were in some instances enveloped in leather or encased in gold, and worn as amulets round the neck.

Many of the stones of the garden wall before me are covered over with a thin coating of vegetation of various hues and forms. The tints from Nature's palette have been applied with wonderful skill; the warmer and more vivid hues gradually blending with the grey and neutral ones. By this means, the harsh, artificial aspect of the wall has disappeared, and an air of natural beauty has been imparted to it, exquisitely harmonizing with the white trunks of the birch-trees, the green flower-sprinkled bank of the streamlet, and the blue cloud-flecked softness of the over-arching sky. Instead of disfiguring, it now adorns the landscape, and the eye rests upon its mottled, softly-rounded sides and top with unwearied pleasure. It affords an illustration of the common truth, that there are no distinct lines of demarcation, no harsh, abrupt objects allowed in nature. Even man's work must come under this law; and wherever Nature has the power, she brings back the human structure to her own bosom, and, while dismantling and disintegrating it, clothes it with a living garniture of beauty, such as no art of man can imitate. The farmer may keep the meadow or cornfield distinct from the surrounding scene, heavy with uniform greenness, or ugly with the discordant glare of yellow weeds; but as soon as Nature obtains the control of it, when out of cultivation, she brings it into harmony with the landscape by carefully spreading her wild flowers over it in such a way as to restore the proper balance of colour. As the earth is rounded into one great whole, so all its objects are connected with each other, not merely by laws of structure and dependence, but also by close æsthetic relations. The rock, decked with moss, lichen, and fern, shades in sympathy of hue and outline with the verdure of wood and meadow around it; the mountain and the ocean melt on their farthest limits into the blue of the sky; the river and the lake do not preserve the distinctness of a separate element, but blend with the solid land, by mirroring its scenery on their tranquil bosom; and the very atmosphere itself, by its purple clouds on the horizon, raising the eye gradually and insensibly from the dull, tangible earth to the transparent heavens, becomes a part of the landscape instead of the mere empty space that surrounds it.

While this picturesque effect of the wall is admired, the objects which produce it are very generally overlooked. If carefully examined, however, they will be found very interesting, both on account of their peculiarities of structure and the associations connected with them. Almost every stone is made venerable, as also the adjoining fruit-trees and espaliers, with the grey rosettes of that commonest of all lichens, the stone parmelia. This plant used to be extensively employed by the Highlanders in dyeing woollen stuffs of a dirty purple, or rather reddish-brown, colour. By the Arabian physicians it was administered under the name of *âchnen*, for purifying the blood; and it was also an ingredient in the celebrated *unguentum armarium*, or sympathetic ointment, which was supposed to cure wounds if the weapon that inflicted them were smeared with it, without any application to the wounds themselves. Besides this lichen, the ointment consisted of human fat, human blood, linseed oil, turpentine, and Armenian bole, mixed together in various proportions. A present of the prescription for this precious mess was made by Paracelsus, about the year 1530, to the Emperor Maximilian, by whom it was greatly valued. Much was written, in the medical treatises of the time, both for and against the efficacy of such applications; and, in an age when prescriptions as a rule were founded upon some real or fancied resemblance between the remedy and the disease, the stone parmelia was an object of great importance. It is now sold by the London herbalists solely for the use of bird-stuffers, who line the inside of their cases and decorate the branches of the miniature trees upon which the birds perch with it. There are also numerous specimens on the wall of the yellow parmelia, no less renowned than its congener in the annals of medicine as an astringent and febrifuge. By Dr. Sander, in 1815, it was successfully administered as a substitute for Peruvian bark in intermittent fevers; the great Haller recommended its use as a tonic in diarrhœa and dysentery; and Willemet gave it with success in cases of hæmorrhages and autumnal contagious fluxes. In the arts it is employed at the present day as a dye-stuff, yielding a beautiful golden yellow crystallizable colouring matter, called chrysophanic acid, which is nearly identical with the yellow colouring matter of rhubarb; and, like litmus, it may be used as a test for alkalies, as they invariably communicate to its yellow colouring matter a beautiful red tint. It is the most ornamental of all our lichens. Its bright, golden thallus, spreading in circles two or three inches in diameter, and covered with numerous small orange shields, decks with lavish profusion the rough unmortared walls of the poor man's cottage; and many a rich patch of it may be seen covering the crumbling stones of some hoary castle or long-ruined abbey as with a sunset glory. Growing in a concentric form, when it attains a certain size the central parts begin to decay and disappear, leaving only a narrow circular rim of living vegetable matter. In this manner it covers a whole wall or tree with spreading ripples of growth and decay—analogous to the fairy rings formed by the growth and decay of mushrooms in a grassy field. This yellow wafer of vegetation is attached to the stone by slender white hairs on the under surface, looking like roots, although they do not possess the power of selecting and appropriating the materials of growth peculiar to such organs. We know not by what means lichens derive nourishment. Some species certainly do disintegrate the stones on which they occur, and absorb the chemical and mineral substances which they contain, as is clearly proved when they are analysed. But a far more numerous class are found only on the hardest stones, so closely appressed and level with their surface that they seem to form an integral part of them. In this way they continue for years, ay centuries and ages, unchanged—their matrix as well as their own intense vitality resisting all decay. There are instances of encaustic lichens covering the glaciated surfaces of quartz on the summits of our highest hills, which may probably be reckoned among the oldest of living organisms. Such species can obviously derive no benefit save mere mechanical support from their growing-place, and must procure their nourishment entirely from the atmosphere, and their colouring matter from solar reflection.

The eye of the naturalist, educated by practice to almost microscopic keenness, can discern scattered over the wall numerous other specimens of this singular vegetation, appearing like mere discolorations or weather-stains on the stones. Some are scaly fragments so minute as to require very close inspection to detect them. Others are indefinite films or nebulæ of greyish matter, sprinkled with black dots about the size of a pin's head. Others are granular crusts of a circular form, with a zoned border; and when two or three of them meet together, they do not coalesce and become absorbed into one huge overgrown individual. The frontier of each is strictly preserved by a narrow black border, however it may grow and extend itself, as zealously as that of France or Austria. The law against removing a neighbour's landmark is as strictly enforced in lichen as in human economy. When a stone is covered with a series of these independent lichens, it looks like a miniature map of Germany or America; the zoned patches resembling the states, the black dots the towns, and the lines and cracks in the crust the rivers. There is one species growing on pure quartz, an exquisite piece of natural mosaic of glossy black and primrose yellow, called the geographical lichen from this resemblance.

Several of the stones are sprinkled with a grey, green, or yellow powder, as dry and finely pulverized as quicklime or sulphur. These grains are either the germs of lichens awaiting development, or they are individual vital cells, capable of growing into new plants, in the absence of proper fruit. The pulverulent lichens are always barren, because a strict individualization of each cell is at variance with the regular formation of organic fructification, since in the latter the individuality of the separate cells appears most circumscribed and checked. It is difficult to distinguish these pulverulent masses from the powder of chalk, verdigris, or sulphur; and yet they are endowed with the most persistent vitality, which almost no adverse circumstances can extinguish. The principle of life resides in each of these grains as truly as in the most complicated organism; and, though reduced here to the very simplest expression of which it is capable, it is not divested of its mystery, but on the contrary rendered more wonderful and incomprehensible. A wide and impassable barrier separates these life-particles from the grains of the stone on which they occur, and yet it is very difficult in some cases to distinguish the one from the other. The extreme simplicity of structure displayed by these protophytes is more puzzling to the botanist than any amount of complexity would have been. The rudimentary stages of all the flowerless plants appear in this singular form. The germs of a moss are similar to those of a lichen, and the germs of a lichen to those of a fern or sea-weed. These powdery grains represent the basis from which each separate system of life starts, to recede so widely in the highest forms of each order. The advocates of spontaneous generation or development—for there is essentially little difference between these two theories—have endeavoured to derive from this circumstance a plausible argument in support of their views. They assert that the germs of all cryptogamic plants are not only apparently, but essentially, the same; and that the differences of their after development are owing to accidental circumstances of soil, situation, and other physical conditions. If they happen to fall upon decaying substances, they become fungi; if they are scattered in soil, they become ferns or mosses; if water is the medium in which they are produced, they grow into algæ; and on dry stones and living trees they spread into the flat crusts of lichens. Plausible as this idea looks, it is not borne out by experiment, for the same germs sown in the same soil, exposed to precisely similar conditions, develop one into a moss, another into a lichen, a third into a fungus, and a fourth into a fern; showing clearly that though we cannot discover the difference between their rudimentary germs, a real distinction does nevertheless exist—that the seeds of these minute, insignificant plants are in reality as different from each other, as the seed of an apple-tree is different from that of a pine or palm. The developments of nature are not regulated by accidents and caprices; they are the results of fixed, predetermined laws, operating in every part of every living organism, from the commencement of its growth to the end of its life-history. And the similarity which we find between them is not the consequence of a lineal descent of one from another, but only a feature of the same grand plan of construction: the resemblance is not the result of anything in these forms themselves; it is a purely intellectual relation of plan. With this small piece of granite before me, then, what solemn and far-reaching questions are connected! Geologists of the Plutonian and Neptunian schools have keenly contested the mode of its formation; while arguments drawn from the living particles of vegetation on its surface have been advanced in support of the "development" and "origin of species" theories. Could we explain the mysteries locked up in this little stone, we should be furnished with a key to the mysteries of the universe.

When the powdery lichens occur in large quantities, they give a very picturesque effect to rocks, trees, and buildings. The trunks and branches of trees in the outskirts of large towns are covered with a green powder, fostered by the impurity of the air; a similar substance is also produced in damp, low-lying woods, where the trees are so densely crowded as to prevent proper ventilation and free admission of light. In Roslin Chapel, near Edinburgh, the curious effect of the rich carvings of the walls and pillars is greatly enhanced by a species of *Lepraria*, of a deep verdigris colour, covering them with the utmost profusion. It gives an appearance of hoary antiquity to the structure, and is the genuine hue of poetry and romance. On boarded buildings, old palings, and walls may be sometimes seen a greyish film sprinkled with very red particles, turning yellow if rubbed, and exhaling when moistened a very perceptible odour of violets; from which circumstance it has obtained the name of *Lepraria Jolithus*. Linnæus met with it frequently in his tour through Œland and East Gothland, covering the stones by the roadside with a blood-red pigment. It also spreads over the wet stones of St. Winifred's Well in North Wales, and is supposed to be the blood of the martyred saint—a superstition which, like the dark stain in the floor of Holyrood Palace, one has not the heart to disturb. I know not if others have realized the sentiment, but I have often felt as if I could willingly have given up all the knowledge I possess of the structure and history of these obscure productions, in exchange for the power of being able to look upon them with the childish wonder which in early unscientific days they inspired. There is an air of mystery and obscurity about them peculiarly fascinating, which it is not desirable to dispel by the garish light of technical knowledge. Each one of them seemed a self-discovered treasure of childhood, as much my own as if God had made it on purpose and presented it to me; and it was ever a part of my joy to think that I had found something which no one else knew or had seen before, and that I could bestow upon it pet names of my own. They were links connecting me with an unseen, unexplored world, where the marvellous was quite natural—parts of the scenery amid which elves and fairies, and all the denizens of the heaven that lies about us in our infancy, lived. So many strange things, the existence of which we never suspected, then presented themselves to our notice every day, that nothing seemed impossible or supernatural. Precise limits have now fixed for us the extent of our domain, and we know everything within it. "First a slight line, then a fence, then a wall; then the wall will rise, will shut in the man, will form a prison, and to get out of it he must have wings. But around the child neither walls nor fences—a boundless extent, all irridescent with brilliant colours." How full to the brim with beauty were the flower-cups that were on a level with the eyes of the little botanist. We men have outgrown the flower and all its mystical loveliness!

It is among the mosses of the wall, however, that the richest harvest of beauty and interest may be gathered. Long have my mingled wonder and admiration been given to these tiny forms of vegetable life—beautiful in every situation—spreading on the floor of ancient forests, yielding carpets that "steal all noises from the foot," and over which the golden sunbeams chase each other in waves of light and shade throughout the long summer day—throwing over the decaying tree and the mouldering ruin a veil of delicate beauty—honoured everywhere of God to perform a most important though unnoticed part in this great creation. Well do I remember the bright July afternoon when their wonderful structure and peculiarities were first unveiled to me by one long since dead, whose cultivated eye saw strange loveliness in things which others idly passed, and whose simple warm heart was ever alive to the mute appeals of humblest wild flower or tiniest moss. There was opened up to me that day a new world of hitherto undreamt-of beauty and intellectual delight; in the structural details of the moss which illustrated the lesson I got a glimpse of some deeper aspect of the Divine character than mere intelligence. Methought I saw Him not as the mere contriver or designer, but in His own loving nature, having His tender mercies over all His works—displaying care for helplessness and minuteness—care for beauty in the works of nature, irrespective of final ends or utilitarian purposes. Small as the object before me was, I was impressed —in the wonder of its structure, at once a means and an end, beautiful in itself and performing its beautiful uses in nature—not with the limited ingenuity of a finite, but with the wisdom and love of an Infinite Spirit. To that one unforgotten lesson, improved by much study of these little objects alike in the closet and in the field, I owe many moments of pure happiness, the memory of which I would not part with for all the costly, painted pleasures, to gather which, as they ripen high on the wall, the world impatiently tramples down things that are far sweeter and more lasting.

A careful search will reveal upwards of a score of mosses on our garden wall, in almost every stage of growth, from a dim film of greenness to radiating plumes spreading over the stones, and cushion-like tufts projecting out of the crevices, and crowned with a forest of pink fruit-covered stems. One is amazed at the exuberance of life displayed on so small and unpromising a surface. It gives us a more graphic idea than we commonly possess of the vast and varied resources of creation. Though so much alike in their general appearance as to be often confounded by a superficial eye, all these species are truly distinct; and when closely examined exhibit very marked and striking differences. They are not slightly varying expressions and modifications of the same Divine idea; but rather different ideas of creative thought. Each of them stands for a separate revelation of the Infinite Mind; and the fact that the same plan of construction, the same type of character, runs through them all, only indicates that there is everywhere, in the minutest as well as most conspicuous parts of creation, an undeviating regard to unity and harmony.

Prominent among these mosses are the curious little tortulas, found abundantly on every old wall —when there is sufficient moisture and shade—but loving especially the rude stone gable and thatched roof of the Highland cottage, covering them with deep cushions of verdure till the whole structure appears more like a work of nature than man's handiwork. I have always great pleasure in looking at this tribe of mosses through a lens. The leaves are beautifully transparent and reticulated, and readily revive, when scorched and shrivelled by the sunshine, under the first shower of rain. The most noticeable thing about the tortulas is the curious fringe which covers the mouth of the seed-vessel. In all the species, of which there are about fourteen in this country, the fringe is twisted in different ways like the wick of a candle. This peculiarity may be easily seen by the naked eye, as it projects considerably beyond the fruit-vessel, and is of a lighter colour; but the microscope reveals it in all its beauty. It is a wide departure from the ordinary type, according to which the teeth of the fruit-vessel are made to lock into each other, and thus form a wheel-like lid, composed of separate spokes, which fill up the aperture. The great length of the teeth in the tortulas prevents this arrangement of them; their tops are therefore twisted, as the farmer twists the sheaves at the top of his wheat-stack, so as to keep out the rain; and this plan seems to answer the purpose as effectually as the normal one. Some of the tortula tufts are of a pale reddish colour, as if withered by old age, or scorched by the sun. This peculiar blight extends in a circular form from the centre to the circumference of a tuft, where filmy grey textures, like fragments of a spider's web interweaving among the leaves, proclaim the presence of an obscure fungus, in whose deadly embrace the moss has perished. Thus even the humblest kinds of life are preyed upon by others still humbler in the scale; and perhaps there is no self-existent organic structure in nature. Besides this parasite, there are other species of life nourished by these tufts. If one of them be saturated with moisture, and a drop squeezed out upon a glass, and placed under a good microscope, the muddy liquid will be found swarming with animalculæ, little animated cells, wandering with electric activity amid the endless mazes of the strange forest-vegetation; and among them there is sure to be one or more lordly Rotiferas, lengthening and contracting their transparent bodies as they glide rapidly out of view, or halting a moment to protrude and whirl their wheel-like ciliæ in the process of feeding—the most interesting of microscopic spectacles.

One of the commonest of the mosses on the wall is the little grey *Grimmia;* looking, with its brown capsules nestling among the leaves, like tiny round cushions stuck full of pins. The nerves of the leaves project beyond the point, and give an appearance of hoariness to the plant, in fine keeping with the antique character of the wall. This moss grows on the barest and hardest surfaces—on granite and trap rocks, where not a particle of soil can lodge; and yet every cushion of it rests comfortably upon a considerable quantity of earth carefully gathered within its leaves, which must have been blown there as dust by the wind, or disintegrated by its own roots from the substance of the rock. Our garden wall displays two or three tiny tufts of a curious moss occurring not very frequently on moist shady walls built with lime. It is called the Extinguisher moss, because the cover of the fruit-vessel is exactly like the extinguisher of a candle, or the calyx of the yellow garden *Escholtzia.* We have also a few specimens, in the more retired crevices, of the *Bartramia,* or apple-moss—one of the loveliest of all the species—with its bright green hairy cushions and round capsules, like fairy apples. It fruits most abundantly in spring, appearing in its full beauty when the primrose makes mimic sunshine on the brae, and the cuckoo gives an air of enchantment to the hazel copse. A subalpine species, it is somewhat uncommon in lowland districts; but it would be well worth while to grow it in a fernery. Its Latin name appropriately perpetuates the memory of John Bartram—one of the most devoted of American naturalists—a simple farmer and self-taught, yet a man of great and varied attainments, concealed by a too modest and retiring disposition. Linnæus pronounced him "the greatest natural botanist in the world." It is a touching thing to think of the names of scientific men, great in their own generation, being linked with such obscure and fragile memorials. They have passed away, and with them the memory of all they achieved; and nothing now speaks of them save a little plant, of which not one in a thousand has ever heard, and which only a few naturalists see at rare intervals. There are hundreds of such names in the nomenclature of botany, worthy of a prominent and enduring remembrance, of which almost nothing more is known than this simple association. It is the plant alone that perpetuates them—history and epitaph all in one—like the chronology of the antediluvian patriarchs; and we are apt to smile when we read of the gratification which the illustrious Linnæus felt when the little bell-flowered *Linnœa*, pride of the Swedish woods, was baptized with his name—regarding it as a pledge of immortality; for if there had been nothing but this floral link to connect his memory with future ages, very few would have known that there ever was such a man.

The line of turf along the top of the wall is a perfect Lilliputian garden. It bears a bright and interesting succession of plants from January to December. The little lichens and mosses claim exclusive possession of it during the winter months; for these simple hardy forms of life are most luxuriant when the weather is most severe; they are the first to come to any spot, and the last to leave it—growing through sunshine and gloom with meek and unruffled serenity. There are whole colonies of that most social of all cryptogams, the hair moss, looking, with their stiff and rigid leaves, like a forest of miniature aloes; preserving during summer and autumn a uniform dull green appearance, but breaking out in spring into a multitude of little cups of a brilliant crimson colour, nestling among the uppermost leaves, and rivalling in beauty the gayest blossoms of flowers. Hardly less interesting are the scores of cup-lichens—holding up in their mealy sulphur-coloured goblets dewy offerings to the sun, like vegetable Ganymedes. And the lover of the curious will be sure to notice the livid leathery leaves of the dog lichen, tipped with brown shields like finger-nails, that grow redder in the piercing Christmas cold—bringing us back in fancy to the days of Dr. Mead, the famous physician and friend of Pope, Bentham, and Newton, by whom it was first brought into notice as a remedy for hydrophobia. These and numerous other minute forms, too obscure to mention, may be seen all the year round; and dim though the sunbeams of winter may be, they search them out in their hidden nooks, and stimulate them to life and energy, and the glow of sunrise or sunset, that sets a mountain range on fire, rests lovingly on the smallest moss or lichen, intimating that it too has its place and its relations in this wide universe. When the first mild days of early spring come, the *Draba*, or whitlow-grass, puts forth its tiny white flowers, and greets the returning warmth, when there is not a daisy in the meadow, or a single golden blossom on the whinny hill-side. Then follows a bright array of chance wild flowers, wayward adventurers, whose seeds the winds have wafted or the birds have dropped upon this elevated site, their colours deepening as the season advances—old thyme, ever new, hanging down in fragrant festoons of purple; yellow bedstraw—the Chrysohoë of flowers—like masses of golden foam, scenting the breeze with honey sweetness, and ever murmurous with bees;

chimes of blue-bells hanging from the wall as from a belfry, and tolling with their rich peal of bells—which the soul alone can hear—the knell of the departing flowers. A fringe of soft meadow-grass covers the turf, whose silken greenness forms the ground colour on which these bright patterns are embroidered; while its silvery panicles hang in all their airy grace over the flowers, like gossamer veils, greatly enhancing their beauty. That patch of grass softens no human footfall of care, but it is refreshing to the eye, and the robin rests upon it, as it pours out its low sweet chant, according well with the sere leaves and the dim stillness of autumn, the calm decay of earth, and the peace divine of heaven. I love, in the silent eve, when there is scarcely a breath in the garden, and the sunset is flushing the flowers and purpling the hills, to sit near that richly-decorated wall, in full view of its autumn flowers, smiling on the lap of death, for ever perishing, but immortal—joys that have come down to us pure and unstained from Eden, and amid a world of progress will be transmitted without a single leaf being changed to the latest generation. Looking at them, and feeling to the full the beauty and wonder of the world, I enjoy all that the coming centuries can bestow upon the wisest and the happiest of our race. Voiceless though they are, they have a secret power to thrill my heart to its very core. They speak of hope and love, bright as their own hue, and vague as their perfume; they speak of the mystery of human life, its beautiful blossoming and its sudden fading; and, more than all, they speak of Him, who, holy, harmless, undefiled, and separate from sinners, found on earth most congenial fellowship with these emblems of purity and innocence; whose favourite resort was the garden of Gethsemane; whose lesson of faith and trust in Providence was illustrated by the growth of the lilies; and who, at last—as the Rose of Sharon and the Lily of the Valley—was laid in a sepulchre in a garden, leaving behind there a sweet and lasting perfume, which makes the grave to all who fall asleep in Him a bed of sweet and refreshing rest.

CHAPTER IV.

A RAMBLE THROUGH NORWAY, THE CRADLE OF THE HIGHLAND FLORA.

HAVING exhausted the botany of the British hills, I was anxious to study our Alpine plants in their original centre of distribution, to compare the forms which they present under different conditions of soil, climate, exposure, &c.; and thus ascertain the value of the distinctions, not merely among the species reputed to be doubtful, but also among those commonly considered to be well-established. For this purpose I undertook, two summers ago, along with some friends, a short tour in Norway. I went first to Denmark, a country which holds out many inducements to the botanist, and presents peculiar facilities for exploring, the expense of travelling being extremely moderate, the language interposing but few difficulties to one who knows broad Scotch and low German, and the plants of the woods and marshes being singularly attractive and interesting. The "Flora Danica," a splendid work of some twenty volumes, exquisitely illustrated, contains a great many species that are common in this country; while it gives an admirable idea of the character of the Scandinavian vegetation as a whole.

The beech-woods are the most remarkable feature of Denmark. They clothe the whole face of the country, except the cultivated parts, giving it a soft, rich, languid look, exceedingly pleasing to the eye of one accustomed to the bleak hills and pine-woods of the Scottish Highlands. Hardly any other tree besides the beech is seen in these forests now; but it was not always so. Denmark is a palimpsest of three distinct layers of arboreal vegetation. In the lowest stratum of the bogs trunks and other portions of Scotch fir-trees are found; above this layer is a distinctly-marked stratum in which nothing but remains of oak occur; while the surface of the country is covered with flourishing beech-forests. These changes in the character of the woods indicate corresponding changes in the character of the climate; for the oak is now a rare tree in the country, and the Scotch fir is never seen, being unsuited to the altered circumstances. The age of these extinct forests is a much-disputed question. Worsa, the able director of the Museum of Northern Antiquities in Copenhagen, showed me several very interesting human relics dug from these deposits, which must have belonged to the Bronze period, and probably dated no further back than the time of Abraham. The antiquity of the Danish Kjökkenmödings, like that of the Swiss Pfahlbauten, has been unduly stretched. But apart from archæological speculations, which in such spots are irresistibly suggested, nothing can be more delightful than a ramble among the beech-woods in the neighbourhood of Copenhagen on a hot summer day. The shadows are so cool and deep; the belts of golden light that lie across the greensward at every opening among the trees are so bright and sunny; the far-stretching vistas so mysterious and seductive to the imagination; and the trunks and branches of the beeches so smooth, round, and well-filled, and so covered with heavy masses of beautiful transparent foliage, that you feel as if in an enchanted place. You think longingly of the long-ago times when an English county merited its beautiful poetical name of "Buckinghamshire,"—"the home of the beech-trees;" beech being the modern form of the old Teutonic *buck* or *buch*.[1] From a rising ground,

---

[1] There are some interesting peculiarities in the geographical distribution of the beech. It is the tree which ascends highest on the Apennines, forming large forests immediately below the zone of the Alpine plants. On Gran Sasso d'Italia, the loftiest peak of the range, it flourishes luxuriantly at a height of 6,000 feet above the Adriatic; not far from the line of perpetual snow. On other mountain chains it is the birch or the pine which ascends the highest, and adjoins the zone of the Alpine flora. In Norway the beech is unknown, save in the extreme south and in the plains; while in the Alps it occurs only in the lower valleys. On the Apennines it ascends several thousand feet above the region in which corn can be cultivated, and where man lives permanently; and yet corn is grown as far north as Lapland, and a few patches of it ripen even at Hammerfest. The reason of this is that the beech is affected by the heat of the whole year, while the corn depends upon the summer heat. The heat of summer in the Arctic circle is much greater than at a height of 6,000

through a break in the forest, you catch a glimpse of the blue waters of the Sound flashing in the sunlight, with white, spirit-like sails flitting to and fro over its placid bosom: you thus feel that the place is haunted for ever by harmonies of winds and waves—visited by delicate influences from sea and land. Occasionally, at the end of a vista among the trees, a solitary deer may be seen feeding, or pausing to gaze at the stranger, and gliding silent as a shadow into the remoter recesses. The ground is everywhere enamelled with the wild flowers which we see in our own woodlands; and every sight and sound are so homelike that it is difficult to realize the idea that one is in a foreign land. I saw large patches of the yellow wood-anemone (*A. Ranunculoides*) and of the *guul fugls melk* (yellow bird's milk), *Ornithogalum luteum*, but they were both past flowering. When in full bloom, in spring, they make the woods quite a California. In this primitive country almost every plant is known to the peasant, and

feet in Italy, while the cold of winter is much more severe. Altitude and latitude, which correspond so far as herbaceous plants depending upon summer heat are concerned, do not correspond so far as trees which depend upon a certain degree of heat all the year round are concerned. It is because of the somewhat uniform annual temperature of the Antarctic regions—less warm in summer and less cold in winter—that the Evergreen beech (*Fagus Forsteri*) forms the characteristic woods of Terra del Fuego, and the Antarctic beech grows even farther south in the Antarctic regions, while no species of beech can flourish in the extreme climate of the Arctic regions. The abundance and beauty of the beech in Denmark is doubtless owing to the same cause.

associated with some quaint incident. I was greatly struck with the beauty of the lichens, mosses, and fungi, which grew upon the trunks of the trees, and especially upon the fallen ones, in moist and shady spots. Many of the beeches were sprinkled with the rich yellow powder of the lichen *calicium;* others were covered with the chocolate patches of the tamarisk scale-moss; while on several prostrate trunks I found the curious fungus *Dædalea* growing to an enormous size, and exhibiting on the under side its intricate sinuosities, like a Chinese carving in ivory. I gathered some foreign plants which afford an illustration of the curious way in which the flora of one country finds its way to another. When the statues which Thorwaldsen sent from Rome were unpacked in Copenhagen, several flowers sprang up very soon after in the neighbourhood formerly unknown. It seems that the sculptures were carefully wrapped round with bands of hay from the Campagna, containing the seeds of plants peculiar to Italy. Might not the incident be regarded as typical of Thorwaldsen's own genius, which had grown and been developed in the Eternal City, and at last blossomed in old age in his native place? I observed in moist, rocky dells, among the moss, great quantities of the *Primula farinosa.* The leaves and stalks were powdered with the characteristic lemon-dust, but the beautiful lilac flowers were overpassed, and fruit formed. In early May the market-women come into town, bearing basket-loads of this lovely flower tied in little nosegays. The glens of Lyngby and Ramlösa are covered with it in spring. Many spots in the neighbourhood of Copenhagen in this delicious season are like the Vale of Tempe. Indeed, at any time, nothing can be more soothing to ruffled nerves than the serenity and loveliness of Danish scenery. Doctors should send their patients, jaded and excited by the hurry and over-work of our large towns, to these peaceful drowsy retreats, where the very spirit of repose has made its home and the mere fact of existence is a delight. Denmark is indeed a land where it seems always afternoon; and the lotus-eater can wander day after day among its beech-woods, and never weary of the monotony.

But we tore ourselves away ere the beech-woods had completely bewitched us with their sorceries. More bracing and stimulating work awaited us among the dark fjords, snowy fjelds, and pine-forests of Norway. This country possesses a peculiar interest to a Scotchman, not only because it is the original home of the Highland flora, but chiefly on account of its former intimate connexion with the northern and eastern parts of Scotland Colonies of Norsemen occupied these parts sufficiently long to effect a radical change in the appearance and manners of the primitive inhabitants—transforming the undersized Celt, afraid of the sea, into the bold, adventurous, finely-developed seaman. From this source were derived the fair hair, blue eyes, and straight limbs which characterise a large proportion of our seafaring population, as well as the names so common among them, ending in *son*, Anderson, Henderson, Johnston, Paterson—which are the most frequent at the present day in Norway—and the peculiar terms applied to the Scottish firths, bays, and promontories. With pleasant hopes kindled by these associations, we embarked on Wednesday, 25th June, on board the *Viken*, a Government steamer regularly plying in the postal service between Copenhagen and Christiania. Our passage was a somewhat stormy one among the white waves of the Cattegat. But after we had passed Gottenburg, on the Swedish coast, at which we had called about two o'clock next morning, when the town was buried in profound repose, all the rest of the voyage was calm and beautiful, and there was nothing to mar our high enjoyment of the wonderful intricacy and picturesque shores and islands of the Christiania Fjord. Retiring to rest after leaving Moss glowing with the indescribable hues of a northern sunset, we awoke from a very unrefreshing sleep about six o'clock on Friday morning, and found the steamer quietly moored to the quay of Christiania.

The morning was very bright and sunny. Hastily dressing ourselves and collecting our traps, we stepped ashore, glad enough to exchange the heaving deep for solid earth, and the coffin-like airless berths of the steamer for a limitless supply of fresh air, blowing from the hills of Gamle Norge. A few leisurely porters and drowsy Government officials, blinking in the sun, were lounging about, and neither bustle nor business reminded us that we were standing on the quay of a metropolis. After waiting a while, a custom-house officer condescended to examine our luggage, with his hands in his pockets and a cigar in his mouth; and as we carried no contraband goods, not even a flask of Glenlivet or a canister of "bird's-eye," we were let off very easily, and our crumpled toggery was speedily repacked. We tried two of the hotels which the English are in the habit of frequenting, but fortunately for our purses we found them quite full, and were at last obliged to take refuge in the Hôtel Scandinavie, where we were charged something like native prices, and had no reason to complain either of the fare or the attention. We were told that it was a great gala day in Christiania, a market being held there called St. Hans' Fair, at which timber-merchants from every part of Norway meet to buy and sell wood. We should certainly not have found out this fact ourselves, for the streets appeared to us exceedingly quiet and deserted, only two or three people at long intervals walking very slowly along the rough pavement, smoking the eternal cigar, and wearing an air of leisureliness and repose, as if they were the heirs expectant of time, most provoking to a fidgety and active Englishman. Most of the population seemed to have congregated in our hotel, overflowing bedrooms, stairs, and lobbies, treading on each other's toes, distracting the hapless waiters by their multifarious commands, and filling all the air with a confused clattering of unknown tongues.

Christiania does not awaken much admiration in a stranger's mind. It is a very small city to be a capital, and none of the buildings are either ancient or imposing; most of the picturesque log-houses that used to exist having been destroyed by fire and replaced by plain brick buildings without any architectural features. There are few shops, and these generally small and shabby, dealing in miscellaneous ware, like a druggist's emporium in an English country village. The best places of business are in the Kirke-gaden; but the Norwegians have so little skill and taste in displaying their goods in the windows, that even the finest shops present but a poor appearance outside. Some beautiful pieces of filigree silver, of native metal and manufacture, may be purchased in this quarter, as well as very ingenious specimens of Norwegian carving, an art in which the inhabitants, especially of Telemarken, rival the Swiss and Germans; but the prices to English visitors are generally very high. The people in their intercourse with one another and in their business transactions outdo the Parisians themselves in politeness. Their hats are more frequently in their hands than on their heads; and the magnificent sweep of the bow with which one grocer acknowledges the presence of another in the street always elicited my unqualified admiration. The free and independent British tourist, who persists in wearing his hat alike under the dome of St. Peter's at Rome and in a Christiania curiosity shop, suffers immensely by the comparison; and a blush of guilt rose to my own cheek on more than one occasion, when, in momentary forgetfulness that I was not at home, I entered a shop with my hat on, and was recalled to painful consciousness by the significant pantomime of the shopkeeper. If slaves cannot breathe in England, Quakers certainly could not exist in Norway.

The only buildings that are at all handsome are the Storthing or House of Commons, where the Parliament of Norway meets once every three years to transact during three months a very large amount of gossip and a very small amount of business; the university, with its library and museums; and the palace of the king, situated on a commanding eminence above the town, and surrounded by gardens kept in a very slovenly style, the walks of which are a favourite promenade of the citizens in the cool of the evening. We visited this palace. It was guarded by a solitary shabbily-dressed sentinel, who paced slowly backwards and forwards with a slouching gait, stopping every ten minutes to rub a lucifer match against the wall of the building and light a penny cigar. We asked him if we could get admittance, and he pointed out to us a small side door, at which we knocked. A tall, fat, good-natured woman appeared, and, conducting us through a series of underground passages, brought us up to the principal entrance-hall, from whence we followed her over the whole building. We found the palace of Charles XV. very similar to other palaces. There were great rooms of state with much *bizarre* gilding and little comfort; and there were small rooms with little gilding and great snugness. The private apartments of the king, queen, and heir to the throne were very plainly furnished; and the bedrooms in which royalty takes the sleep that, according to the Turkish proverb, makes pashas of us all, had small curtainless beds like sofas, and couches draped with a very threadbare-looking tartan of the clan M'Tavish. I suppose the descendant of Bernadotte, on the same etymological principle that Donizetti was proved to be the Italianized form of the Celtic Donald Izzet, was a ninety-second cousin of some Highland family, and therefore took the tartan. The view of the fjord and surrounding country which we obtained from the leaden roof was truly magnificent, and decidedly the most regal thing about the palace. A wide expanse of sea stretched out before us, calm and blue as an inland lake, studded with innumerable islands, covered with ships and boats sailing in every direction, each floating double, ship and shadow, in the transparent water, and bounded in the distance by an irregular grouping of picturesque hills, which gave the fjord a varied outline like the Lake of the Four Cantons in Switzerland. Immediately below was the old romantic castle of Aggershuus, situated on a bold promontory of the sea, and adorned with fine avenues of linden-trees along the ramparts. This castle was besieged and taken by the redoubtable Charles XII. of Sweden, and now contains the regalia and the state records of Norway. Close to the old town rose up the hill of Egeberg, richly cultivated and wooded to the top, and commanding an extensive prospect on every side. Westwards, the white tower of Oscar's Hall —a summer residence of the King of Norway, and containing a fine series of Tidemand's paintings— peeped out with picturesque effect from the midst of a perfect nest of foliage, while the landscape in that direction was perfected by the snow-capped mountains of Valders and Telemarken visible in the far background. Everywhere there were rich woods, not only of pine and fir, but of deciduous trees, elm, plane, ash, lilacs, and laburnums, growing in the utmost luxuriance. On every side there were cultivated fields, picturesque groups of rocks, gleaming waters, rugged hills, and elegant villas embosomed among fruit-trees and flowering shrubs. I know of no town that has so many country-houses scattered around it; and it would be difficult to say which of them is most beautifully situated. Each has its own separate view, its own woody knoll, and cultivated field, and rocky islet, and vista of the fjord. And this wondrous combination of art and nature makes the environs of Christiania quite a fairy scene. The sky, too, was so mellow and blue, the air so clear and sunny, and the colouring of the landscape so intense and glowing, that I almost fancied myself in Italy instead of on the 60th degree of north latitude—in the parallel of the Shetland Islands. The only scenery which the view from the palace suggested to me was the southern extremity of the Lake of Geneva, looking across the outskirts of the town to the Jura mountains; but the comparison is greatly in favour of Christiania.

We paid a visit, as in duty bound, to Mr. Bennett, who is the great authority on matters Norwegian to all Englishmen—reverenced by them almost as much as Murray or Bradshaw. He lives amid a curious collection of novels, " Leisure Hours," old broken-down carrioles, silver drinking-cups, and a lot of mixed pickles and Worcester sauce; the last supposed to be absolutely essential to the existence of the British tourist in Norway. He acts in so many capacities that he must be a kind of universal genius, being antiquary, librarian, purveyor, custom-house agent, *Deus ex machinâ* of the Christiania Carriole Company, and last, not least, churchwarden and collector of subscriptions for the English chapel in town. He has done, I have heard, many kind and disinterested acts to strangers introducing themselves to him; and he has been repaid in too many instances by dishonesty and ingratitude. We did not need the aid of his topographical knowledge, however, as we had previously sketched out our tour with remarkable fulness, and were determined to adhere to the programme in every particular. We therefore contented ourselves with buying from him the last edition of the "Lomme reiseroute," or Government road-book, and a translation or commentary upon it in English, called "Bennett's Handbook," both of which we found exceedingly useful, indeed indispensable, on the journey; for an appeal to the prices of posting marked in the "Lomme reiseroute" was never disputed by the station-house keepers, and it saved much loss of temper and waste of time in haggling about payment.

Having seen all that was to be seen in the way of curiosities about Christiania—which certainly was not much—we took out tickets on the following Saturday for a short ride of forty-five miles on one of the only two railways in all Norway, as far as Eidsvold, the Norwegian Runnymede. The railway was constructed by British navvies, and the railway carriages were made in Birmingham. Proud of our country's universal services to humanity, we rolled along at the rate of eight miles an hour, over a broken country of pine-woods, lakes, and rocky foregrounds, till we came at last to the scene of the Convention which framed the present admirable constitution of Norway. Here we embarked on the Miosen Lake in a steamer, boasting the funny name of *Skibladner*, so called from Odin's magical pocket-ship. This lake is the largest in Norway, being 63 miles long and about 7 broad. It is very highly praised by the Norwegians, and the scenery on its banks is considered the finest they have. This is owing, however, to the same law of contrast which made the Swiss peasant say to the Dutchman, when told that Holland had not a single mountain, "Ah! yours must be a fine country." The Norwegians have so little arable land, and such an overwhelming preponderance of huge barren mountains and rocky plateaux, that the scarcer article as usual is most valued, and the profitable is preferred to the picturesque. The proportion of soil under culture, or capable of being cultivated, to the entire extent of the country is not more than one to one hundred; while upwards of forty per cent. of the surface of the southern half exceeds 3,000 feet above the level of the sea. We were a good deal disappointed in the scenery, having heard it compared to Lake Como, with which it has not a single feature in common. It is a fine sheet of water for boating purposes and for the transport of timber, through many rafts of which the steamer in some places fought its way; but the shores at the lower extremity are banks of bare clay, crowned on the top with a few miserable birches, and farther up the land around it lies low, and is thickly dotted with red wooden farmhouses and variegated by potato and corn fields; while the hills beyond are of no great elevation, and are covered with interminable forests of sombre pines, which produce a melancholy impression by their extreme monotony—especially when, as is usually the case, the sky overhead is grey and cloudy. We landed at about half-past nine at night at a pretty large village at the head of the lake, called Lillehammer, amid the silver splendours of a very singular sunset. This village is built upon an elevated terrace, a considerable distance above the shore of the lake, and commands a most extensive view. Bare brown mountains sprinkled with patches of snow gird the horizon, and give an air of Alpine loneliness and wildness to a landscape that would otherwise have been too rich and luxuriant. The houses, which are all built of wood, are very clean-looking, and neatly painted in pale colours of pink, yellow, and green, which are frequently renewed. Many of them are surrounded by gardens and orchards, or embosomed among clumps of white-stemmed birches and purple lilacs. In every window of every house, even the poorest, are pots of the most brilliant flowers, roses, calceolarias, verbenas, geraniums, petunias, and many other plants, which one would not expect to see in such a latitude. They are most carefully and skilfully tended; and even in a duke's conservatory such perfectly-formed and gorgeous blossoms are rare. The love of flowers is quite a passion with the Norwegians. Go where you will—in the large towns and in the loneliest parts of the country—you will find the windows of the houses filled with the choicest plants, even the humblest making an effort to grow something green and brightly-coloured, that may remind them of a world of beauty beyond their own bleak hills. A philosopher like him who made out murder to be one of the fine arts, who is fond of tracing the final causes of human phenomena, might find the reason of this universal floral mania an interesting subject of speculation. It may be caused by the love of contrast; the eye seeking relief in the bright colours of roses, geraniums, and calceolarias, from the extreme monotony of the green pine-forests and dark brown fjelds. At any rate, the red and other gay colours of the dwelling-houses, and the Oriental brilliancy of the costumes of the people, may fairly, I think, be attributed to this cause.

We spent the Sunday in the village, and had the privilege of worshipping in a little Lutheran church not far from Hamar's inn, where we stayed. It was a welcome rest to body and soul. The day was very beautiful, calm and soft, with wandering gleams of sunshine breaking through the grey clouds, and illuminating here and there the shadowy pine-woods and the cornfields with a more vivid greenness. The lake lay still as a mirror below, with belts of light and shade crossing its bosom, and yellow timber rafts lying motionless along its shores. At intervals the mellow monotone of the cuckoo, whose Norwegian name *gowk* is the same as the Scotch, came from the far-off pine-woods; nearer at hand, in the green fields, the corn-craik uttered its harsh cry; while the roar of the numerous waterfalls of the Mesna, a powerful stream that flows through the village down into the lake, sounded very loud in the universal Sabbath stillness. After dinner I walked up the heights into the shadows of the pine-woods. I sat down with my Bible in a very peaceful and beautiful sanctuary of nature. Before me, a fine cascade gleamed white through the trees, and filled the wood with its psalm of praise. Around me, the red trunks of the pines stretched away into endless vistas of green loneliness and odorous gloom. The ground everywhere was carpeted with rich and rare mosses—cushions of that loveliest species, the ostrich plume feather moss, and tufts of the *Lycopodium annotinum.* There was one splendid lichen peculiar to Norway and the Arctic regions, called *Nephroma arctica*, which I saw in this wood for the first time. It formed an immense rosette, upwards of a foot in diameter, of primrose yellow lobes, their under side tipped like finger-nails with the rich chocolate-coloured fructification. It was really a most beautiful plant, spreading over the ground everywhere, and would have been more in keeping with the luxuriance of a tropical forest than with the monotony of a Norwegian pine-wood. The mossy carpet was starred with fragile wood-sorrels and white coral-like bilberry blossoms. I read the first chapter of Revelation, and mused upon it, until I too had a revelation of Jesus Christ in my Patmos; saw the hairs of His head in the white flowers around me, and His eyes and feet in the flaming sunset that burned through the trees; and heard His voice in the cataract, like the sound of many waters, and felt, like "Aurora Leigh"—

> "No lily-muffled hum of a summer bee,
> But finds some coupling with the spinning stars;
> No pebble at your feet, but proves a sphere;
> No chaffinch, but implies the cherubim;
> . . . . Earth's crammed with heaven,
> And every common bush afire with God;
> But only he who sees takes off his shoes."

We started from Lillehammer early on Monday morning, through the valley of Gudbrandsdal, to Molde, a distance of nearly 200 miles, in a north-western direction. Fortunately, there was at the village a four-wheeled English carriage that had brought a party from Molde to the Miosen Lake, and now waited to be brought back to its owner. We got the carriage free on the condition of paying for the horses; and this arrangement materially lessened the expense of the journey, as well as added greatly to the comfort of the ladies of the party. We formed a somewhat imposing procession as we passed through the village, and attracted a considerable share of attention from the inhabitants. The vehicle which contained my friend and myself was what is called a *stolkjerre*, or double carriole. It was simply a square unpainted box, mounted on two wheels, without springs, and furnished with long shafts and a hard board laid across for a seat. It held us both tightly jammed; free to turn our heads round, but not our bodies. The animal did not reflect much credit upon his species, and his accoutrements consisted of a most complicated and ragged system of grey cord and old leather. Altogether it was a sorry turn-out, and it would require a considerable amount of moral courage to drive through London in it. But the villagers thought it rather grand than otherwise; at least the boys did not run after us, and a few peasants actually doffed their caps. On we sped, seeing the rich hilly scenery in glimpses through the dust of our chariot wheels, with frequent and loud exclamations of "Oh!" as the machine made a rougher jolt than usual. After about an hour and a half's drive, the carriage suddenly disappeared up a by-road. But we, absorbed in conversation, or in looking at the scenery, had not noticed this movement; and thinking the carriage was ahead, though out of sight, drove confidently onwards at full speed. We were alarmed when we had gone a few hundred yards by hearing shouts in very energetic Norwegian—meaning probably "Stop thief!"—and seeing half-a-dozen fellows bounding rapidly towards us through the brushwood above the road. One of them came forward, and, mounting on our vehicle, without a word of explanation seized hold of our reins, and drove us back prisoners up a side-path till we came to a cluster of wooden houses, where we halted. It seems that we had arrived at the first of the series of stations placed for the convenience of travellers at distances of about one Norwegian, or seven English miles, through the whole length of the Gudbrandsdal valley. The horse and machine we had brought with us from Lillehammer must here be changed for a fresh horse and machine, and the boy who had accompanied the horses of the carriage had to take them back along with our equipage. Hence the alarm of the natives at our ignorant escapade. They thought that we were going to run off with our magnificent dog-cart, and sell the whole affair for a large sum at Molde. Of course, had they known that we were clergymen, they would not have insulted us and excited themselves by cherishing such fears; but there was nothing in our appearance to indicate our profession, and I suppose our faces, apart from our professional habiliments, were not accepted as conclusive evidence of our honesty.

I must here pause a little to give an idea of the mode of travelling in Norway, as this is a convenient halting-place for the purpose. There are no stage coaches or diligences, for the people very seldom travel, and then only on pressing business. The most common and characteristic vehicle of the country is called a carriole, shaped somewhat like an old-fashioned gig. It has no springs, but the shafts are very long and slender, and the wheels very large, so that its motion is far from being uncomfortable. It carries only one person, who has to drive with his feet nearly on a level with his nose, and a boy sitting behind on the portmanteau, amalgamating its contents, whose duty it is for an exceedingly small *drikkepenge* or gratuity to take back the horse and machine. Owing to this arrangement, a large party must go in a long file of carriages like a funeral procession. The Norwegian horses are all small, cream-coloured, and remarkably docile and sure-footed, so that the most timid lady or the youngest child might safely drive them down the steepest gradients at full speed. The roads are made by Government: but each proprietor along the highway has to keep a certain portion of it in good working order, this portion being regulated according to the size and value of the property through which it passes. Painted wooden poles are placed at certain intervals along the road, inscribed with the name of the person who has to keep that part of it in order, and the number of yards or *alen* entrusted to his supervision. You can, therefore, form a pretty good idea of the wealth or poverty of any neighbourhood through which you travel by the greater or less distances of road thus distributed to the owners of land. At regular intervals of seven or eight English miles—as already observed—there are placed station-houses, where fresh horses and conveyances may be had, as well as lodging and entertainment for man and beast. These stations are either fast or slow stations. At the fast stations a number of horses and carrioles are kept regularly, ready for the convenience of travellers; so that you ought not to be detained on your journey more than half an hour. A printed Government-book is kept at each of these stations, where the traveller writes down his name, the number of horses and carriages he requires, the place he has come from, and his destination, as well as any complaint he may have to make on the score of carelessness or detention. Such complaints are inquired into regularly by a Government inspector, and redressed as far as possible. Some of the remarks made in the column of complaints by Englishmen are very amusing. There was one English name which we found in the road-book of every station, coupled with some depreciating remark upon the scenery, the manners of the people, the nature and price of food, &c. &c. Nothing seemed to please his jaundiced eye and bilious stomach. Doing the journey post-haste, a detention of ten minutes in changing his horse and carriage at a new station was a most exaggerated offence. Desirous of making a profit of his tour, by spending less for travelling and keep together than his ordinary personal expenses would have cost at home, the charge of fivepence for a cup of coffee with solid accompaniments was considered most exorbitant. Here the people were excessively disobliging, and he was half-starved upon strong-smelling *gamle ost* (old cheese), parchment-like *fladbrod*, of which nearly an acre is required to satisfy an ordinary appetite, and butter that looked like railway grease; there the eggs were all rotten, there were no toothpicks, and the landlord was an extortionate Jew. With a slight variation upon the same lively tune he went from place to place. Fortunately, as English was not the language of the country, his Parthian shafts did not wound so severely as he intended. On the contrary, it was amusing to see the conscious pride with which his ill-natured remarks were pointed out to us by more than one innkeeper, who imagined in the innocence of his heart that they could not be anything else than highly laudatory. We were glad to see that others of our countrymen, following in the wake of Mr. Smith, had reversed his decision, and by their genial and hearty commendation of many things that were really excellent saved Englishmen from the imputation—which they too often justify abroad—of being a nation of grumblers. And while I am on this subject I may as well mention that very great harm is done to the peasantry by the thoughtless and indiscriminate lavishness on the one hand, and the excessive meanness and stinginess on the other, of our countrymen. The simple-hearted people cannot understand the inconsistency; and Norway promises, if the same demoralizing system continues to be pursued as at present, to be a second edition of Switzerland and the Rhine—a result which every one who knows and can appreciate the primitive straightforwardness, the genuine kindness, and honest independence of the Norwegians must deplore.

At the slow stations the peasants of the neighbourhood are obliged by turns to supply the traveller with a horse and conveyance; and, unless he sends a *forbud* or messenger before him to apprise the people of the exact time of his coming, he may have to wait several hours while the horse is being caught on the hills. Of course, should the traveller disappoint the station-keeper, either by delay or by failing to appear altogether, compensation must be given. We had no experience of these slow stations, for all the stations on the route we took were fast, so that we got on very swiftly and pleasantly. We met no English travellers all the time; and our claims for horses and conveyances were never brought into competition with those of others. Some of the stations are poorly furnished, and very scantily supplied with provisions. You may riot in Goshen-like plenty to-day, and to-morrow be reduced to *fladbrod* and porridge. The traveller who passes in the morning may fare sumptuously upon reindeer-venison, ptarmigan, and salmon; while he who comes late in the day may have to content himself with polishing the bones and gathering up the fragments which his more fortunate predecessor has left. In some quarters the innkeepers shift so frequently that no dependence for two successive years can be placed upon Murray's certificate of character; and we ourselves found the best entertainment, the greatest attention, and the most moderate charges, in places marked dangerous on account of the very opposite qualities. Many of the stations are filthy, and uninhabitable by any one more refined than a Laplander, swarming with F sharps and B flats. Indeed, the king of the fleas keeps his court—not at Tiberias, as travellers say—but at a Norwegian station-house of the worst class. We, however, were either more fortunate than the great bulk of tourists, or our bodies were unusually pachydermatous, for in no case were we tormented during the night watches, and generally the larder was well supplied with salmon, trout, beefsteaks, and eggs. The price of accommodation was ridiculously low—at least when compared with the bill of a Highland hotel. We had a magnificent supper, a capital bed, and a breakfast consisting of more than six dishes of a very solid character, at the first station we halted at, and the cost of the whole was only 1s. 10½d. for each. The price of accommodation, as well as the charge for horses and conveyances, is fixed by Government tariff, but the innkeepers invariably ask more from Englishmen, as they imagine that every native of these islands who travels in their country must be an embryo Rothschild. The usual rate of keep per day is a specie-dollar,—that is, 4s. 6d. of our money; and the day's travelling expenses, along with keep, unless you go enormous distances at a stretch, should very rarely exceed an average of 10s. The station-house keepers are a very respectable class of men, usually. They are often landed proprietors or justices of the peace, and only set themselves out for the entertainment and transport of travellers because they are obliged to do so by Government. Indeed, this innkeeping and posting business is a tax, and they pay it as we pay income-tax, with something like a grudge. They must, therefore, be treated with civility, and in some instances with very considerable respect. A Norwegian innkeeper, if ordered about like a Highland Sandie M'Tonald, would considerably astonish the traveller guilty of such boldness.

But to return from this digression, necessary to explain our mode of travel, to the route itself. The road through the Gudbrandsdal is the regular postal route from Christiania to Throndhjem, and is therefore the most frequented and the best known part of the country. And yet the people are almost as unsophisticated as in the remotest districts. They crowded around us at the different stations, questioned us on all sorts of subjects, and carefully examined our dress and luggage. The ladies of our party were especial objects of curiosity to the women. Their ornaments and watches were tenderly touched, and greatly admired. Hands were lifted up in amazement at the strange wonders which glimpses of foreign boots and petticoats disclosed. An air cushion inflated for their benefit, and placed on the carriage seat, and then sat upon by an adventurous Dutch-built dame, elicited shouts of merriment. A few presents of pins, buttons, and Birmingham trinkets made them insist on shaking hands with us all round, a proof of friendship which, owing to the general prevalence of that touch of nature which makes Norway and Scotland kin, the ladies were somewhat shy of accepting. The flaxen-haired cherubs had a revelation of a higher world than the common world of *fladbrod* and porridge—a foretaste of Valhalla itself—in the unknown delights of English comfits and lollipops; though I am not sure that it was really kind in us thus to awaken capacities and educate senses which, after a momentary fruition of bliss, must thenceforward be craving for the unattainable and "the unconditioned."

After passing several stations, and accomplishing nearly fifty miles, we arrived late in the evening at Listad, near the picturesque and ancient church of Ringebo, where we stayed all night. About half a mile distant from the station-house, a wild gorge between micaceous cliffs is formed by the Vaalen Elv, a large torrent that flows from the mountains on the left into the Lougen. Here, on the white vellum-like bark of the birch-trees, I gathered for the first time in Norway great quantities of that most lovely lichen, the *Cetraria juniperina*, in full fructification. The thallus is richly frilled, and of a most vivid yellow colour, contrasting beautifully with the broad shields of deep chocolate brown borne on the extremities of the lobes. Expanded by the recent rain, this lichen covered with its shaggy tufts all the trunks and branches of the trees, and imparted to the wood a very singular appearance. The mossy ground was also tesselated by large patches of the *Cetraria nivalis*, a snowy scolloped lichen growing in erect rigid tufts, which in this country is only found on the extreme summits of the Cairngorm mountains. The scenery of the Gudbrandsdal valley is praised in the most exaggerated terms by Murray. He says that it affords a series of the finest landscapes in the world, and that it is doubtful whether any other river can show such a constant succession of beautiful views as the Lougen, which flows through it. The valley is indeed remarkable for its length, being 168 English miles long; and the greater part of it is richly cultivated, with pine-clad hills rising on either side, but almost never picturesque in outline, or assuming an Alpine character. It is in fact a mere trough across one of the most massive and featureless mountain chains in Norway, bounded on both sides by comparatively uniform and level background. The great peaks retire behind the sky-line so as to be completely invisible; there are no distant prospects, none of those charming lateral vistas caused by interlacing mountains, which reveal enough only to stimulate the imagination, and solicit it onward to grander scenes beyond. Even in the wildest and most romantic parts of the route, which are considered to be the entrance of the valley between Lillehammer and Moshuus, and the Pass of Rusten, between Laurgaard and Braendhaugen, the view is either exhausted altogether, or, as in passing up Loch Katrine to the west, the eye sees out through the romantic to the tame and flat beyond; thus greatly impairing the impression which such a spot ought to produce. There are many landscapes in the Highlands quite equal, if not superior, to those of the Gudbrandsdal valley. Owing to the peculiar conformation of the mountains, the really splendid scenery of Norway is confined to the fjords of the west coast.

We were greatly charmed with the river Lougen, which, always very broad and deep, expands here and there into chains of lakes—some of which, like the Lake of Losna, are navigable for large vessels. Indeed, for upwards of twenty miles, between Moshuus and Listad, the journey used to be accomplished by a steamer, which has now been withdrawn. Some very fine cataracts occur in the course of the river; and the roar of the immense body of water, broken up into snow-white masses contrasting beautifully with its uniformly rich green colour elsewhere, combined with the picturesqueness of its lofty banks adorned with hanging woods of pine and birch, produce a profound impression. At the Pass of Rusten especially the river is truly sublime, forcing its way through a narrow gateway in the mountains, which approach each other so closely that the road has been cut out of the living rock. It is a fearful place, of which the Pass of Killiecrankie can give one no idea; and we drove shudderingly through it on the brink of precipices overhanging the deep foaming linns of the river. On the gneissic rocks through which the road was cut I observed an immense quantity of a very rare lichen called *Gyrophora murina*, which is included in the list of British lichens on the authority of specimens found on St. Vincent's Rock by a Mr. Dare, but which has not since been seen there, or indeed anywhere else in this country. It consists of a single roundish, crumpled, concave leaf, from an inch to an inch and a half in diameter, attached by a central disk to its growing-place. Its upper side is of a dark ash colour, passing into dark brown on the edges; the under side being of a deep black, covered with minute shagreen-like roughness, interspersed with scattered fibres. The rocks in this locality were completely blackened with it; and were thus harmonized with the profound gloom of the spot. Norway is the head-quarters of this tribe of lichens; which are also common on our highest Highland mountains. On the bare arid rocks behind Christiansand occurs that most singular member of the family, the *Umbilicaria pustulata*, like large ragged patches of dark brown parchment, covered with warts or pustular elevations of the whole surface of the thallus. Below the fortress of Bergenhuus, that guards the harbour of Bergen, I noticed it growing in immense profusion, giving a very shaggy look to the rocks. Common on the granite of Devonshire, and especially on Dartmoor, in Scotland it is only found near the head of Loch Sligachan in Skye. On this route we saw no villages cosily grouped round a church, whose spire is conspicuous from afar. The churches are lonely buildings, few and far between, and the names crowded so thickly on Munch's admirable map indicate mere farmhouses with their steadings, called a *gaard*, equivalent to the Scottish *toun*. This isolation and dispersion of the houses over a wide area is a singular feature in Norwegian landscapes, and arises from the fact that almost every head of a family is the proprietor of the land on which he dwells. It gives, as Professor Forbes has remarked, a dreary interminable aspect to a journey, like that of a book unrelieved by subdivision into chapters, where we are at least invited to halt, though at liberty to proceed.

Next day, before coming to the gorge of Rusten, we passed the cleft of Kringelen, where Colonel Sinclair, nephew of the Earl of Caithness, and his regiment of Scotch mercenaries, were massacred by an ambush of the peasants in 1612. Sinclair offered his services to Gustavus Adolphus, King of Sweden, who was then at war with Norway and Denmark. Landing from Scotland at Molde, he marched through Romsdal, intending to cross the uplands of Norway to the frontiers of Sweden, laying waste the country as he passed with fire and sword, and committing many acts of remorseless cruelty. Exasperated to the utmost fury, and unable to contend with Sinclair in open fight, a band of 500 peasants adopted the same expedient as that recorded in the Tyrolese war of independence. Having collected an enormous quantity of rocks and stones on the brow of the hill immediately above the pathway leading through the narrow defile of Kringelen, they awaited the signal of a young man who had undertaken to guide Sinclair to this spot. No sooner were the devoted troops fairly underneath, and the signal given, than the fatal avalanche descended, burying them under the huge pile, so that only a few escaped. An affecting incident in connexion with this tragic event is commonly told to the traveller. A Norwegian lady in the neighbourhood, hearing that Mrs. Sinclair was with her husband, sent her own lover, to whom she was to be married next day, to protect her from insult; but Mrs. Sinclair, mistaking his intentions, drew a pistol from her bosom, and shot him dead on the spot. It is said that Mrs. Sinclair, a young and beautiful woman, was most devotedly attached to her husband, whom she followed across the sea disguised in male attire, and did not reveal herself until the arrival of the troops in Norway, when she could not be sent home. The dalesmen are never tired of reciting the praises of their valorous countrymen on this occasion. An inscription on a pillar by the roadside marks the scene of the massacre, and tells how "the peasants, among whom dwell honour, virtue, and all that earns praise, brake the Scotch to pieces like a potter's vessel." In the peasants' huts, matchlocks, broadswords, powder-flasks, and other relics of the regiment are shown to tourists with much patriotic enthusiasm. There is a Norwegian ballad entitled "Herr Sinclair's Vise af Storm," sung by almost every native, of the end of which the following is a free translation:—

> " Strike home, ye valiant Northmen all!
> Was the dalesmen's answering cry;
> And fast the Scottish warriors fall,
> And in their gore they lie.

> "The raven flapped his jet black wing
>   As he mangled the face of the slain;
> And Scottish maids a dirge may sing
>   For the lovers they'll ne'er see again.
>
> "No one of the fourteen hundred men
>   E'er returned to his home to tell
> What peril awaits the foe in each glen,
>   Where the stalwart Northmen dwell.
>
> "A pillar stands where our foemen lie,
>   In deadly fight o'erthrown;
> And foul fall the Northman whose heart beats not high
>   When he looks on that old grey stone."

The natives, as in this ballad, try to prove that the slaughter of the Scotch was not a treacherous massacre, but the result of a brave hand-to-hand encounter. And they will not believe that Scotchmen care very little for the fate of Sinclair and his mercenaries, of whom not one in a thousand has ever heard. We certainly did not blush for our country when we surveyed the wild scene.

After passing through the dark gorge of Rustenberg the road gradually ascends, until, at last, an elevation of 1,800 feet above the level of the sea has been attained. The scenery in consequence becomes bleaker and less wooded; the spruce and pine gradually giving place to the birch, which here forms the principal tree—and, as usual, has a whiter and cleaner trunk and brighter foliage in proportion to the altitude.[1] The cultivation of

---

[1] It is interesting to notice that in all probability the name of the *birch* comes from the Sanscrit word *bhoorja*, applied to the laminated bark of an Indian birch (*Betula Bhojpatra*) used for writing and

corn and potatoes is merged in that of grass and hay; and the fields, which look dry and parched, are irrigated by means of wooden troughs, in which water is led down, often for long distances, from the mountains. The air feels keener and more bracing; patches of snow appear in the shady hollows far down the mountain sides on our left; and the landscape assumes a wilder and more Alpine character. At Braendhaugen the road is very sandy; this part of the valley, called Lessoe, which is purely pastoral, having evidently been once the bottom of an extensive glacier lake. Great banks of clay, scantily covered with grass, and presenting a peculiarly bleak grey appearance, rise up on the right-hand side of the river. This feature continues uninterruptedly to Dombaas, and the soil is so loose and sandy that the steep sides of the road are covered with withered patches of artificial turf fastened by wooden nails to prevent them slipping. It is very disagreeable travelling along this part of the route in dry weather, owing to the clouds of dust raised by the vehicles. Following immediately behind the carriage—for our spirited horse could not be kept back—we were

---

ornamental purposes, like the paper-birch of North America. If this be so, it affords a striking proof of the theory that the ancestors of the present races of Europe migrated westwards from Central Asia. The bestowal of the name of an Indian birch upon a similar tree peculiar to the northern latitudes of Europe is as curious in its way as the bestowal of Saracen names, such as Mischebel, Al-al-'Ain, derived from the natural objects of the Arabian desert, upon the mountains and glaciers of Switzerland during the Moorish invasion.

nearly suffocated. Our clothes were as white as a miller's, and the scenery appeared to us all the harsher on account of the scanty glimpses we obtained of it, and the irritation of the gritty particles in our eyes. At Braendhaugen the good old lady who keeps the station showed us the silver cup presented to her by the Queen of Norway and Sweden; but my recollection of this stage hangs chiefly upon a pair of magnificent reindeer antlers nailed above the door, indicating that reindeer venison is occasionally found here.

We were very tired after the long day's journey; the heat and dust had been very oppressive; and, for my own part, the jolting on a cushionless seat had made me so sore and tender that I could scarcely walk or sit. At eight o'clock at night we arrived at the mountain station of Toftemoen. Here we expected to stay all night; but a party from Throndhjem had sent on a *forbud* and secured all the available accommodation, and we had therefore to go on to the next station, where we could get quarters. We were glad, however, to rest a little and get some refreshment at Toftemoen. This is a very ancient place, and famous in the sagas. It is one of the mountain stations which have the privilege of immunity from taxes, and appears to be one of the most comfortable resting-places in Norway. The proprietor is Mr. Tofte, well known throughout the whole country. He is the lineal descendant of Harold Häarfager, the first King of all Norway, and, in consequence of

Odin, the mythological Hercules of the North. The family are exceedingly proud of their birth, and take precedence of all the other proprietors at church and market. They have never been known for many generations to marry out of their own family—the result being that the present owner of the name is a simpleton, and his eldest son nearly a dwarf. This descendant of kings and representative of the oldest family in Europe unharnessed our horses for us like any common stable-boy. I treated him with considerable deference—though whether he was more impressed by my manner or my attempts at Norwegian I cannot say. But, in return, he showed me the principal rooms in his house, which contain many curious old cabinets, and a broad slate table on which the present King of Norway and Sweden dined on his way to be crowned at Throndhjem. I saw the king's autograph, which he had scratched with a knife at one corner of the table. Tofte told me, with an air of considerable self-importance, of the dignified reception which he had given to the king; and related that, when the king wished to bring out his silver for dinner, he replied that he had as much silver in his house as would suffice to dine a much larger party than the king's. This was no idle boast, for I never saw in a private person's dwelling such a vast quantity of massive silver articles, evidently heirlooms, dating, some of them, many centuries back. Besides being possessed of the bluest of blue blood, Tofte is a wealthy landed proprietor, a member of the Storthing or House of Commons, and a justice of the peace. This did not prevent him, however, from charging us a higher price than we had paid anywhere else for the entertainment we had at his house. He presented me with his photograph taken at Christiania, dressed very stiffly and uncomfortably in Sunday clothes. The face is intensely Scotch, with a peculiar look of combined simplicity and shrewdness.

The rest of our journey that night was not very pleasant, and it was past eleven o'clock when we arrived at the telegraphic station of Dombaas. All was quiet and still; the people apparently having gone to bed, and sunk into the first deep sleep. Though so late at night, there was no darkness. You could read the smallest print with the utmost distinctness; and but for the stillness of nature, and an indefinable feeling of mellowness and tenderness in the air, you might imagine it to be noon instead of midnight. The long bright Norwegian twilight is inexpressibly beautiful. The earth sleeps, but her heart waketh; the golden tints of the departing day still linger on the distant hills; and a light, soft and sweet as the smile of an infant in its first slumber, fills all the sky, and you would think that the dawn had returned, only that the glory is in the west instead of in the east. Nothing reminds you of darkness and sleep but the rich liquid lustre of Venus hanging near the pale blue horizon, like a silver lamp let down out of heaven by an unseen hand, and flecking a little shadowy pathway of light upon every exposed sheet of water. The long daylight is very favourable to the growth of vegetation, plants growing in the night as well as in the day in the short but ardent summer.. But the stimulus of perpetual solar light is peculiarly trying to the nervous system of those who are not accustomed to it. It prevents proper repose and banishes sleep. I never felt before how needful darkness is for the welfare of our bodies and minds. I longed for night; but the farther north we went, the farther we were fleeing from it, until at last, when we reached the most northern point of our tour, the sun set for only one hour and a half. Consequently, the heat of the day never cooled down, and accumulated until it became almost unendurable at last. Truly for a most wise and beneficent purpose did God make light and create darkness. "Light is sweet, and it is a pleasant thing to the eyes to behold the sun." But darkness is also sweet; it is the nurse of nature's kind restorer, balmy sleep; and without the tender drawing round us of its curtains, the weary eyelid will not close, and the jaded nerves will not be soothed to refreshing rest. Not till the everlasting day break, and the shadows flee away, and the Lord Himself shall be our light and our God our glory, can we do without the cloud in the sunshine, the shade of sorrow in the bright light of joy, and the curtain of night for the deepening of the sleep which God gives His beloved.

We had considerable difficulty in arousing the people from their slumbers, but at last we succeeded in obtaining the services of a blithesome lass, who speedily extemporized beds for us, and made us as comfortable as possible on such short notice. The beds in Norway, I may mention, are all procrustean; a kind of domestic guillotine invented for the purpose of amputating the superfluous length of Englishmen's legs. The Norwegians are a tall race, but I suppose they lie doubled up in bed like the letter V, the os coccygis touching the footboard, and the feet and head keeping loving company on the same pillow. Though not above the average height, my own unfortunate limbs were hanging exposed over the footboard; the down quilt lay in all its rotundity in my arms like a nightmare of some monster baby; and, while sleeping uneasily in this awkward posture, I dreamt that I had been metamorphosed somehow into a waterfall, and was flowing in white masses of foam, and with a considerable murmur, over very hard and slippery rocks. Next morning we felt the air a good deal colder, for we were now at an elevation of upwards of 2,000 feet above the level of the sea. The scenery of the place was bare treeless upland, very sparingly cultivated. The road to Throndhjem passed in a series of ups and downs over monotonous brown hills to our right; while the highway to Molde lay far down in an equally featureless valley to our left. A few hillocks here and there broke the level surface, covered with grey boulders, and clothed, instead of heather, which is somewhat rare in Norway, with crowberry and arbutus bushes. The lovely large blue-bells of the Menziesia peeped up everywhere among the familiar moorland vegetation; the Andromeda displayed its rich crimson blossoms on every dry knoll; while the clayey banks were brightened and beautified exceedingly with multitudes of the fairy Scottish primrose, whose sulphury leaves and tiny purple flowers are the ornament of the Caithness cliffs, but proceed no farther south in this country. There was an air of inexpressible loneliness about the place; the stillness being broken only by the feeble bleat of a few sheep and goats—as diminutive, though full-grown, as lambs and kids—and the tinkle of the bells suspended round the necks of the no less Lilliputian cattle. A few pigs ran about, as thin as greyhounds; and the Alpine vegetation, as well as the small size of animal life, testified to the ungenial character of the climate. The coolness of the air was very pleasant to us, roasted as we had been so long in the confined valley; but it must be a very trying thing to live at this elevated station in winter. Storms must blow over its shelterless fields with unexampled fury, and the snow drift in huge masses around it. The short black December day will be like the frown of Odin, and every wild night lit up by the magical radiance of the Aurora Borealis will be a Walpurgis-Nacht. Woe to the traveller who is then obliged to cross the Dovrefjeld!

After leaving Dombaas, the scenery became exceedingly tame and uninteresting. Huge featureless mountains of gneiss scantily clothed with brown moorish vegetation enclosed a dreary valley covered with straggling pines. The road at first passed over a desolate height among stunted firs and junipers—where immense cairns of stones blackened with tripe-de-roche lichens and Alpine mosses everywhere encumbered the ground. The pastures here were very bare and stony. Large tufts of the aconite or monkshood, peculiar to Alpine pastures, spread over them as thickly as the yellow rag-weed spreads over a fallow field in England. The sheep and cows were miserably thin and ill-fed. It was a poverty-stricken region, sadly contrasting with the rich Gudbrandsdal and the fertile Romsdal, between which it lay. Most of the houses were rude hovels of the most primitive construction. We noticed that a considerable number of the birches by the roadside had a broad ring of black round their white stems. The bark had been stripped off to cover the roofs of the houses; shingles or turf being laid above. This birch bark has a very pleasant smell, and is besides very durable and quite impervious to moisture. The walls were made of squared trunks of trees, ingeniously dovetailed at the corners, with layers of sphagnum or bog moss inserted between each log, in order to keep out the cold. From these squalid abodes crowds of bareheaded, barefooted children in fluttering picturesque rags rushed out as we passed by, clamorous for alms, following us for long distances with their importunities. As the road in this locality was crossed at frequent intervals by gates, separating the numerous small farms from each other, this circumstance was taken advantage of in earning an honest penny. No sooner did our carrioles appear in sight than a boy would rush out from a house, with three pieces of rag floating behind him, and run with headlong speed along the road to open the nearest gate for us. Frequently, however, his hopes of a skilling or two were disappointed by the forethought of a longer-headed comrade, who had stationed himself at the gate in readiness to open it at once to the expected travellers. In such cases, we always rewarded the honest labour of the legs, and not the slothful cunning of the brain. The Lovgen at this part of the route passes through several lakes, the largest called Lesje Vand, and the smallest Lesje Vâerks Vand, which is 2,078 feet above the level of the sea. Here a rare and curious phenomenon in physical geography may be seen. The river Lovgen, whose course we had been following for upwards of 200 miles from Lillehammer, issues from the last-mentioned lake on the south-east and flows through the Miosen Lake to the Christiania Fjord; while the Ravma issues from the other extremity and, flowing to the north-west through the valley of Romsdal, falls into the Molde Fjord. The whole of Southern Norway is thus surrounded by water, and converted into an island.

Passing a miserable place called Holager, we arrived very early in the day at Holseth, a very clean and comfortable station. As we had resolved to remain here over the night, I embraced the opportunity of ascending one of the Dovrefjeld mountains, upwards of 4,000 feet high, immediately in front of the inn. The first part of the ascent was exceedingly arduous, leading through a tangled maze of junipers and dwarf birches (*Betula nana*), creeping over loose fragments of rocks, and forming the underwood of a splendid forest of Scotch firs. I was delighted to find here the *Pyrola uniflora*, perfuming the air with the delicious fragrance of its large erect snow-white blossom. In boggy places grew a remarkably beautiful and stately species of "rattle" (*Pedicularis sceptrum Caroli*), called by the people *Karl's skefter*. It is peculiarly a Lapland plant, and I was astonished to find it so far south. It is upwards of three feet high, the upper half being a spike of golden flowers. Rearing its lofty head above the grass, it looks like a royal sceptre, and is a great ornament to the wood. In the same moist localities I also found the stately *Angelica archangelica*, whose pungent aromatic stems, called *myrstut*, are highly prized by the Norwegians for their stomachic properties, and eagerly gathered wherever they can find it. Above the forest region, the mountain, though very much steeper, was less encumbered with shrubs, and therefore more easily climbed. The most abrupt declivities of the Norwegian hills are invariably on the western side, the eastern side having a gradual inclination, while the summits consist of broad flat tablelands. Owing to this feature, the various zones of vegetation do not rise above one another as in the Alps and other mountain chains, but rather lie side by side; so that you may travel several days on slightly rising ground through the region of the firs; for several days more through the zone of the birch; and for an equal length of time through the belt of the Alpine plants, before the snow-covered ridge is attained. The botany of Norway, therefore, over very wide spaces of mountainous territory, is somewhat monotonous, presenting none of those quick transitions which form the charm of Alpine exploration, and rendering botanizing a work of time and great fatigue. The season of my visit happened to be a late one—the previous winter having been unprecedentedly severe and protracted. The side of the hill was therefore covered with brown and matted grass, smoothly pressed by the snow that had very recently lain upon it; and on the top there were great snow-wreaths, over which I walked with considerable difficulty. Few of the Alpine plants had yet begun to flower; but in many places exposed to the sun I observed enormous patches in full bloom of the Alpine azalea. The foliage could not be seen for the multitude of rosy flowers. In this country we see it only in little tufts or fragments, which, however beautiful, give no idea of its exquisite loveliness when growing, as on the Norwegian mountains, in solid masses of colour almost acres in extent. Its beauty was greatly enhanced by a setting of reindeer-lichen, which whitened the ground everywhere with its snow-white coral-like tufts. It is with lichens as with Alpine plants; they increase in beauty and luxuriance the higher the altitude or latitude. Every one is familiar with the reindeer-moss of our own moorlands; but the variety that grows on the mountains of Norway and the plains of Lapland is far lovelier, forming dense and much-divided tufts of snowy purity and exquisite shape. The rosy flowers of the azalea gleaming among these lichens looked like rubies or garnets set round with filigree work of frosted silver or carved ivory.

Every dry stony knoll on the hill was covered with the compact cushion-like masses of the Greenland saxifrage, with dense tufts of the Alpine Lychnis (which in this country, as already mentioned, is only found in one locality), or with carpets of mossy campion. Here and there, in marshy places, the rare *Andromeda hypnoides* formed bright green mossy tufts, from whence arose a profusion of slender hair-like crimson stalks, each bearing a single white bell-shaped blossom. Side by side with it grew the *Pedicularis lapponica,* whose soft yellow blossoms formed a pleasing contrast; and the globular snow-white heads of the rare cotton-grass (*Eriophorum Schevchzeri*). Owing to the lateness of the season, the *Anemone vernalis* was still in flower on the sunny slopes, distinguished by its shaggy calyx clothed with brownish silky hairs, and its large white blossoms tinged with purple. But it was among the cryptogamic plants that I gathered the richest harvest of species. The droppings of horses immediately above the fir-forest were covered with no less than four species of *Splachnum*—that rarest and loveliest genus of mosses, viz. *S. rubrum, luteum, ampulaceum,* and *vasculosum.* The first two are peculiar to Norway and the Arctic circle; and the last two are found on the highest Scottish hills, forming dense cushions of large transparent foliage around springs. It is the peculiarity of this singular tribe of Alpine mosses that they almost all grow on organic substances, such as skulls of sheep and deer; one species having been found on the decayed hat of a traveller who had perished amid the snows of St. Bernard. On the summit of the hill, the ground was covered everywhere with dense erect tufts of *Cornicularia ochroleuca,* and the snowy scolloped *Cetraria nivalis*—lichens which in this country are found very sparingly distributed only on the highest summits of the Cairngorm range. The stems of the former are sulphur-coloured, about half a foot long, repeatedly branched, the ultimate branches tinged with a dark greenish hue, as if a faint foreshadowing of grass. Nothing could exceed its luxuriance in this spot, forming carpets into which the foot sank up to the ankle. The rocks were whitened with the large granulated branchy excrescences of the *Stereocaulon paschale,* a lichen common on our own hills; which is remarkable as being the first trace of vegetation that appears on naked lava, and is therefore very general on Vesuvius, Etna, and Ischia. On this Norwegian plateau we have the exact counterparts of the *tundra* or plains that border the Polar sea, covered almost exclusively with dense masses of the same cryptogamic vegetation, and forming the pastures of innumerable herds of reindeer. As if to increase the resemblance, I found many of the lichen tufts and patches of *Ranunculus glacialis* growing beside the snow, cropped as if reindeer had been feeding there very recently; and fortunately lifting up my eyes, I saw over the shoulder of the hill, about a quarter of a mile off, a herd of about sixty reindeer quietly grazing—one buck with large branching antlers standing as sentinel, and the light-coloured does and fawns collected in the centre of the group. It was a romantic sight, and would have delighted a sportsman's heart. In a little while they were apparently alarmed by something, and rushed away, till they were mere specks on the snow of the opposite hill. The reindeer are fast disappearing from the southern mountains of Norway, where they used to be exceedingly numerous, and retreating to the northern parts; and this is owing, not only to the disturbance of their haunts by an increasing number of sportsmen, but also to the gradual amelioration of the climate. It is rare now to see herds of any size farther south than the sixty-third degree.

The Dovrefjeld Mountains are to Norway what the Breadalbane Mountains are to Britain—the finest botanical field in the country. They have been successfully investigated by the late Professor Blytt and his son, the present accomplished curator of the Christiania botanical gardens, with whom I had a pleasant meeting on the Sogne Fjord; but a very large portion remains still to be explored. The greatest variety of rare plants is found about Fogstuen, Jerkin, and Kongsvold. A great number of species peculiar to the Polar circle, and unknown elsewhere in Norway, may be gathered in these places. A large succulent species of moonwort (*Botrychium virginicum*) occurs on the Dovrefjeld, which has a very remarkable geographical range. In Europe it is found only in Norway; but it abounds in many parts of the Southern United States, grows on the Andes of Mexico and on the Raklang Pass in the Himalayas, and is frequent on the mountains of Australia and New Zealand, where it is boiled and eaten by the natives. Like the *Erigeron alpinus* and *Phleum alpinum*—a species of Alpine grass both growing on the Dovrefjeld, on the British mountains, at a great height on the Himalayas, and in the Straits of Magellan and the Falkland Islands—the distribution of this plant over such widely-separated areas is a very puzzling problem.

I do not know whether I was the first who ascended this nameless mountain of the Dovrefjeld, but I gathered a cairn of loose stones which I piled above one another on the highest point, and writing my name, address, and date of visit on a card, enclosed it in the centre for the benefit of future explorers. The view from that elevated spot was truly grand: behind me Snæhattan — long considered the highest hill in Norway—towered up 7,700 feet above a bleak desert plateau; its upper half covered with snow, and, forming an amphitheatre, broken down on one side by great black precipices enclosing true glaciers. Over against me stretched the peaks, pinnacles, and horns of the Langfjeld; while a lofty snow-cone rose stern and solitary on the distant horizon, which I identified as Galdhoppigen, now ascertained to be the highest Norwegian mountain, being nearly 1,000 feet higher than Snæhattan. Westward I saw the fantastic summits of Romsdal, with the sphinx-like form of Storhattan, reposing amid the splendour of golden clouds, and facing the setting sun as if looking over the verge of the earth and peering into another and a brighter world. It was altogether a view peculiar to Norway, with an air of utter desolation and gloomy grandeur. Such vast masses of inorganic matter filled the horizon, that the presence of a little plant beside my feet was cheering — reminding me of the organic chain of sympathy that bound us together. No creature appeared in sight, either on the earth or in the sky. No tinkling of cow-bells or shrill goat-song — sounds elsewhere common — broke the oppressive lifelessness and loneliness of the place. For upwards of an hour I sat on my cairn drinking in the sublime influences of the scene; but the waning light warned me that the day was far spent. In descending I had to traverse a long snow-field as smooth and hard as ice, and lying at a pretty steep angle on the hill-side. I had no sooner stepped upon it than my feet went from under me, and I glissaded with great rapidity down the slope, striking very hard against some birch stumps that protruded out of the snow at the bottom. I was soaked to the skin, and a good deal stunned; but I forgot every bodily discomfort in astonishment at the strange sight which my fall had disclosed. I had noticed before stepping on the snow that the surface was of a curious salmon colour in some places, and covered with fine particles like brick-dust; and now I found that wherever my body had pressed the snow together there was a long crimson streak, as if a creature's blood had been shed there. This was the famous red snow, which is so frequently found in the Arctic regions and on the Alps, produced by an immense multitude of microscopic plants, consisting only of gelatinous cells. Captain Ross on one occasion noticed a snowy ridge extending eight miles in length, tinged with this singular hue to a depth of several feet. Vast masses of it spread over the Apennines in 1818; and it is recorded that in the beginning of this century the vicinity of Belluno and Feltri was covered with rose-coloured snow to the depth of twenty centimètres. The snow is not its natural situation, for it is found, like the *nostoc* and other gelatinous algæ, on moist rocks in this country; but its great tenacity of life enables it not only to preserve its vitality when its germs fall on this ungenial surface, but to grow and propagate itself with the astonishing rapidity of its family, favoured by the heat of the sun and the melting of the snow. Its colour in this country, when growing on rocks, is green; but it has been observed that there is a curious coincidence between a white ground and a red flower; so that its brilliant carmine hue on the snow may be produced by the excess of light reflected by its chilly habitat. Had I not been familiar with this curious phenomenon—having seen it on the Alps—I should have been alarmed, naturally supposing that the crimson streaks had been shed from my own veins by the accident.

At the next station beyond Holseth, called Stueflaaten, the valley of Romsdal fairly begins. From this point the view of grey Alpine peaks, seamed with watercourses, closing in and shutting up the vista to the westward, is very striking, and stimulates the imagination by the thought of grander scenes beyond. The road, recently reconstructed in the most admirable way, winds along by the side of the Rauma. No amount of praise bestowed upon this river can be exaggerated. It is the finest stream in Norway, combining features which are not united in any other river. Its course, though short, is exceedingly varied and turbulent. For twenty miles it has worn its way by the sheer force of its waters through schistose rocks, and formed deep circular basins, narrow channels, and projecting ledges, over and through which it thunders and foams in the wildest manner. The contrasts of colour exhibited by the pale malachite green of its linns, shading into black in the deeper parts, and the snowy whiteness of its cataracts, were very beautiful, and afforded a perpetual feast of delight. Wishing to enjoy the scenery in a calmer and more leisurely way, we walked between Stueflaaten and Ormen, a distance of nine miles. Although the heat was great and rendered exertion of any kind very fatiguing, I never enjoyed any walk so much; my only regret being that it was so short. Every turn of the road opened up a new and grander scene than before—loftier precipices and wilder reaches of the river. Among the innumerable waterfalls of the Rauma—many of which, deeply hidden between perpendicular walls of rock, can only be seen by lying down on the verge of the precipices and gazing over—the finest is the Söndre Slettefoss, a short distance from the road. An enormous body of water is here hurled about forty feet into a long deep cave worn in the rocks, from whence it issues through a rugged gorge fringed with hanging birches. The noise was deafening; and the mists rising up from the abyss clung in wreaths to the black sides of the rocks and tossed the dripping birches in their swirling eddies. It required a considerable amount of courage to stand on the brink and look over into this wild torment of waters. In a little sunny birch-wood beside this waterfall grew in greater profusion than elsewhere a little flower, called *Smilacina bifolia*, peculiar to Norway. It is closely allied to the lily of the valley, having, like it, two broad leaves; but its cream-coloured blossom is smaller, exceedingly delicate, and foam-like. It completely hid the grass with its snowy sheen, and looked as though the foam of the waterfall borne by the wind to the spot had blossomed into flowers. A beautiful species of *Smilacina*, which grows from two to five feet high, and has plaited leaves and crowded panicles of white bell-shaped flowers, is found on the Himalayas. Its young flower-heads, sheathed in tender green leaves, are used as a pot-herb by the natives, under the name of *chokli bi*.

Ormen, the next station, is most picturesquely situated on a rock overhanging the river, which here flows through a very narrow part of the defile. In front is a dense pine-wood; and on the opposite side of the river a large stream flows obliquely down the face of the hill in one long line of white, dividing at last into two parts, and forming a series of waterfalls into the Rauma. A wooden bridge crosses at this point, and gives access to several comfortable *sæters* and rich green pastures. Storhattan rises above the brow of the hill, but is not visible from the station-house, as an extensive table-land of snow intervenes. This isolated mountain, whose sphinx-like form, wherever it is seen, is one of the most striking features in the landscapes of Romsdal, is of great height, and commands from its sharp semicircular summit a vast range of snowy peaks. The ascent, which takes three hours, is very laborious and in some places highly dangerous. The whole of this region presents peculiar attractions to the sportsman, being famous for its game of all kinds. On the mountains reindeer are not unfrequently met; the copses which run up the sides of the valley are the coverts of the *hjerpe*, or hazel-hen, and the *skov-ryper*, or wood-grouse; while, in the pine-woods, the *capercailzie* (called by the natives *stor-fugle* or *big bird*) is sometimes seen, or at least heard, as it makes a startling noise when it is disturbed, in crashing through the branches. The Norwegian squirrel, which differs from our species, is very numerous hereabouts. Like the Alpine hare and ptarmigan, it changes its colour in winter from brown to grey. The winter skin is greatly admired, forming the *petit gris* of commerce, and is much worn by cardinals in Italy. Tracks of bears have occasionally been found at the foot of Storhattan. The day before we passed, broken branches, fresh droppings, and footprints were seen in the copse opposite the station-house, indicating that Bruin had been very recently there. The only place in Norway where one now has a chance of coming in contact with a bear or an elk is in Sâeterdal.

At Fladmark the river flows smoothly between

richly wooded banks of alders and aspens, and here and there a green meadow sprinkled with golden globe flowers and white Alpine bistort. The water is of the loveliest green colour, and so clear and transparent that the mica stones could be seen glittering in the sunlight at the bottom. It was a perpetual baptism of refreshment; while its pleasant murmur marched along with us like the refrain of a song. From this point for fourteen miles we had an endless succession of the most magnificent views of precipices, peaks, and waterfalls. The only place that can be compared to this part of Romsdal is Loch Corruisg in Skye; but it is a very small and insignificant imitation of the tremendous gorge through which we passed. On one side is a series of vertical walls of rock between two and three thousand feet high, with innumerable waterfalls streaming down their sides or leaping sheer down from the top to the bottom, and filling all the air with the confused echoes of their shoutings. At the extremity of this chain of precipices towers up the famous Romsdalhorn, an inaccessible obelisk of granite upwards of 4,000 feet high, seeming quite close wherever one goes, and, like the Matterhorn, changing its shape according to the point of view. On the other side of the gorge are lofty mountains weathered into the most fantastic shapes. One curious point, bearing some resemblance to a monk, is called Martin Luther; and the whole range receives the name of the Troldtinderne or Goblin Peaks. The breadth of the gorge from cliff to cliff may be about two miles, but it does not look a quarter of a mile, owing to the height of the precipices on either side. The Rauma, here a deep, wide river, flows through it, reflecting on its placid bosom the grandeur around. Everywhere the ground is strewn with huge boulders and fragments of rocks; while green verdure and birchwoods struggle up the talus heaps which have crumbled from the weathered peaks above. All the woods by the roadside were covered as thick as they could grow with wild lilies of the valley, bearing a profusion of snowy blossoms, larger and more fragrant even than the garden ones. In the potato and corn fields, growing in great abundance as a common weed, was the beautiful cornel (*Cornus suecica*) with its white corolla and curious eye of black velvet—an Alpine plant which is only found in a few places on our highest Highland hills. The sun was setting when we arrived at the inn of Aak, and a rich crimson glow shone on the snowy pinnacles around, making them look like pyramids of solid fire; while a sky of inexpressible softness and beauty linked the glorified summits together, and gave the whole scene an ethereal look like fairy land. It was a place where the most callous-hearted might worship as in a temple; and when from every birch and lily of the valley rose up on the still evening air a perfume most deliciously subtle and sweet, my senses were fairly intoxicated, and I will not now repeat the extravagant analogies that ran through my brain.

Aak is the most comfortable and delightful place of residence in all Norway. The inn, which is a plain wooden building by the roadside, is kept by Andreas Landmark, a *lensmand*, or justice of the peace, who also owns a large portion of the valley, and the fishing of the Rauma for a mile or two. His wife is said to be a sister of the Bishop of Bergen, and his daughters can speak English very correctly and fluently, especially Laura, who is a most admirable kousekeeper, and attends personally to the wants of the guests. Nowhere does the tourist feel so much at home or fare so well as here; the visitors' book being full of the most glowing praises of the landlord and his daughters. Elsewhere semi-starved on *fladbrod* and that horrible cheese made of sugar and curd which looks like a Bathbrick, or a lump of diachylon, or half-poisoned by the cooking heresies of ignorant peasants, he here revels in all the luxuries of the country properly prepared and served. The table is lavishly supplied with fish and game of various kinds, and wild fruits in the appropriate seasons. As for salmon, for which the Rauma is celebrated, thanks to the successful fishing of two Englishmen who lived at the inn, we got it so often, and in so many forms, that we were in the end perfectly sick of it. We understood, in a way that we never did before, the stipulation of Scotch servants in former times, when about to engage with a new master, that they were not to get salmon oftener than three times a week. There is certainly something in the air of Norway that acts in an extraordinary manner as a stimulant to appetite, for we ourselves found that two hours after a breakfast of the most solid and varied character, which if partaken of in this country would infallibly lead to a bilious attack, and a course of water-gruel for a fortnight, we were quite ready for another meal as substantial. My bedroom, in a separate wing of the house, was a small pigeon-hole of the most primitive kind, approached by a staircase so steep that I had to perform a series of severe gymnastic feats in getting up, and in going down to go backwards, cruelly scarifying my shins. How the chambermaid managed to bring up a tub full of water for ablutionary purposes, without breaking her neck or drowning herself, was a puzzle which I could not solve. But once in, the room was scrupulously neat and clean, and fragrant with freshly-gathered bouquets of lilies of the valley. The garden close by was a delightful retreat in the evening. It was well stocked with culinary vegetables, which were merely in a germinating condition, and the cherry and apple trees were still loaded with blossoms, although it was the beginning of July. The ardent sunshine working night and day, however, would ripen the garden crop in this high latitude quite as soon as in our country.

Saturday after our arrival was an exceedingly sultry day; the thermometer ninety degrees in the shade, and not a breath of wind moving even on the bank of the river. The mosquitoes were very troublesome, adhering so pertinaciously to our clothes that we could not drive them off: one member of our party suffered severely from their bites. This fondness of the mosquito for blood is an inexplicable fact in its history. It is not its natural food, for the insect abounds in places which no warm-blooded animal frequents, and where man is never or rarely seen, and when permitted to suck its fill it turns on its back and remains thus till it dies. This curious point deserves the study of the physiologist. The Norwegian name for this suicidal phlebotomizer is *mouga* or *mouge*, from whence is derived the Scotch word midge, the pest of our summer woods and river-sides. Lying gasping, perspiring, and tormented with heat and mosquitoes, under the shade of the trees, I looked up with longing eyes to the pure white snow-fields of the Goblin Peaks, so suggestive of coolness and vigour. In vain, however, for none of them could be climbed, and the exertion on such a day would be fearful. Across the river, right in front of the inn, is a hill of moderate height, clothed on the lower part with dense scrub, which promised to be easily accessible. It is called " Mid-dag Hill," because the sun appears above its summit at noon, and it is thus a kind of public clock to the neighbourhood. A pathway leads up to the top, and ladies occasionally ascend. This circumstance caused my friend and myself to undervalue the difficulties of the ascent, and refuse the services of a guide. We were not long, however, in finding that we had been too rash and confident in going alone, for we lost the track, which was frequently concealed under huge wreaths of snow, the relics of the past winter, lingering there on account of the lateness of the season, and were surrounded by precipices in every direction. We managed with great difficulty to reach the highest point to which we could venture with safety, which was not more than thirty feet below the real summit. Here the ground, composed of comminuted schist and moistened by the melting of the snow, was carpeted with dense tufts of the beautiful *Diapensia lapponica*, growing side by side with cushions equally dense of the moss campion. The former plant is peculiar to the Alps of Norway and the Arctic circle, and is distinguished by its large white strawberry-like blossom, which is produced so abundantly as almost to hide the foliage. The rosy flowers of the campion were equally abundant, so that together they made a lovely garden in the wilderness. The lily of the valley, though much dwarfed, ascended here to within a hundred feet of the top, wherever there was soil in the crevices of the rocks. On the pure white quartz veins which protruded from the schist grew in immense quantity a black tufted lichen, of extremely rigid habit, called *Cornicularia tristis*, which is one of the most Arctic, Antarctic, and Alpine lichens in the world, being found at the extreme limit of vegetation on the Alps, Himalayas, and Andes, and in north and south latitudes. Over the shoulder of the hill we caught a glimpse of the Romsdalhorn, lifting its giant finger into heaven, as if upbraiding us for our foolhardiness in venturing so near it. It had a peculiar, weird, awful look, like one of the gods of Scandinavian mythology changed into stone, especially when a small wisp of mist—mysteriously formed, for there was not a cloud in the sky—rose up and partially veiled its summit. The view was wonderful, not only in its extent, but also in the peculiarity of its character. Green fjelds sloping down into the green Romsdal Fjord, and hiding in their recesses greener lakes, contrasted in a curious way with snowy mountains, standing out boldly against a deep blue sky. As we descended it was interesting to watch the gradual closing of the boundary line, and the disappearance first of the snowy peaks, and then of the upland lakes, until at last the precipices of Aak confined our horizon. This descent gave us a considerable amount of anxiety, for, unlike the Scottish mountains, which slope down gradually to the valley and reveal their whole outline from the top to the bottom, this hill was exceedingly precipitous, and we could only see at a time about a dozen yards of steep rock below us, terminating abruptly in blank space, terribly suggestive to the imagination. We capped with stones the more prominent rocks by the side of the path as we ascended; but these beacons were of no more use to us in our descent than the crumbs of bread which the boy in the fairy tale dropped on his track, for we got confused with the sameness and gigantic scale of the features of the hill. Our thankfulness and relief, therefore, in reaching the base in safety may be more easily imagined than described. Our gratitude was still further deepened, when, surveying the hill during our evening walk, we noticed how frequently we had come unconsciously to the verge of precipices over which another step forward would have hurled us, to be dashed in pieces more than a thousand feet below.

Sunday was a rainy day, and all the hills were covered to their bases with thick curtains of mist. It was a wild Sinai-like scene. When portions of the mist occasionally thinned away, revealing glimpses of the snow-flecked rocks, so pure and far up, it seemed like vistas opened in heaven—like the vision of Jacob's ladder, with angels ascending and descending. The grand spire of this natural temple, the Romsdalhorn, was completely blotted out of the landscape; but we heard now and then the muffled roar of its avalanches, its awful bell tolling in the darkness. On Monday we left the Romsdal valley with great regret, and embarking on board a steamer calling at Veblungsnæset, we sailed down the fjord amid pine-clad rocks of the most fantastic forms, and islands white with eider ducks, terns, auks, and puffins. At Molde we landed for two hours. From an eminence behind the town, which is of considerable size, and carries on a large trade in fish and timber, we beheld the wonderfully grand and extensive view for which this place is celebrated, rank rising behind rank of lofty snow-peaks, until the last mingled with the white clouds in the distance. Conspicuous in the front row was the Romsdalhorn, the Matterhorn of Norway; beyond was Snæhattan with its silver helmet; and to the south-east the huge fantastic horn of Perpuatind or Skjorten, curved round and covered all over with snow, even on the under curve. In this direction, farther away, were the shattered Aiguilles of the Langfjeld, and the lofty but unknown mountains at the head of the Stor Fjord. No more bewildering array of Alpine peaks crowds upon the eye from the Righi Kulm. It far surpasses, in my estimation, the famous view of the giants of the Oberland from the platform of the Federal Hall at Berne. The Swiss picture lacks the sea, without which no mountain scenery, however grand, can be complete. But the waters of the Molde Fjord, spreading out into a wide island-studded basin, gave an idealistic charm to the vast amphitheatre of mountains rising beyond; and the lights and shades of a sunny day imparted to sea and mountain a witchery of hue and form which made them perfect. We gazed upon the glorious sight with sense and soul stretched to the utmost tension of admiration. The proverb runs, "See the Bay of Naples and die;" but I would say, "See the view from Molde, and have a joy for ever!"

It was eight o'clock at night when we reached Aälesund, a pretty large town, carrying on a considerable trade in codfish with Spain and Italy. It is situated amid a perfect fastness of rocks and water, quite inaccessible except to a Norwegian sailor;

while the views from it of the distant serrated snow-flecked peaks of the Langfjeld are very magnificent. The whole region around is full of the most interesting historical associations. It was the country of the Sea Kings; and from this wild robber's nest they swept down upon the defenceless coasts of England, Scotland, and France. Here are the ruins of the *borg* or castle of the famous Ganger Rolf, the founder of the Duchy of Normandy, and the ancestor of William the Conqueror. We landed in a boat at the quay, and went successively to the two inns in search of beds, but they were both full, owing to a court of justice then sitting. We had therefore to return and sleep on board the steamer. Next day we sailed, amid the same kind of scenery, down the Stor Fjord, calling at the different hamlets on the shores, and at the head of the intricate creeks; and arrived at six o'clock in the evening at the extremity of a long arm of the fjord, where there was a little village called Aähjem, unknown to "Murray." It was a most solitary place—"the world forgetting, by the world forgot." The daughter of the innkeeper had never seen an English lady before. The son, however, a fine smart young man, who spoke a little English, had been to the Paris Exhibition; and we found in the sitting-room the usual souvenirs of French travel. He was looked upon as a great man by the primitive inhabitants; and certainly a more startling contrast could not be found than between the metropolis of fashion and this lonely, far-off Norwegian village. When we landed, the sky from end to end was of molten gold without a single cloud, while the sun trembled in a furnace of dazzling brilliancy above the waters of the fjord, which seemed like a brazen sea. The surrounding mountains were purple with light, and looked as ethereal as clouds; while the universal stillness seemed like the awe and reverence of nature at the great sight. Among moist friable cliffs at a considerable height above the village, decked with starry saxifrages and Alpine alchemilla, I gathered a great many rare cryptogamic plants; and a birch-wood copse at the foot is especially memorable as the spot where I first noticed in Norway the *Linnæa borealis*, afterwards so common and familiar.

On the following morning we took carrioles and drove up a very steep Alpine road, over a mountain plateau, studded with numerous tarns. On the top of the mountain, beside a lake, we saw a *sâeter* or mountain farm, to which the cattle are sent to pasture in spring and summer, under the care of the daughters and female servants of the farmer. Upon these saeters there are houses of very rude construction, and very poorly furnished, in which the tenants live and carry on all their dairy-work. This saeter-life, alone on the mountains for four months in the year, must be very dreary and monotonous. The servants say that they could not endure it, were it not that their lovers come up to see them on the Saturday evenings, when they put on their best dresses and faces, and have a feast of dairy produce and a merry dance. This custom, however, has been formally prohibited by Government, on account of the injury done on such occasions to the game, for the lovers try to kill two birds with one stone. The saeter which we passed was certainly a very lonely place; the pasturage was scanty; and the house a mere hovel of rough unmortared stones, with a hole in the turf roof for a chimney, and another in the wall for a window. The cattle were very small, and wandered about with bells round their necks, making a sweet musical tinkle that increased the loneliness and sadness of the place. It is not wonderful that in such a region should have arisen the strange superstition of the *Huldre*, a mountain spirit who goes forth in the morning with her spectral herd of voiceless and milkless cows, following at a distance the cattle from the *tro*, or fold, when they are driven out to the pastures, and returning with them in the evening. The saeter-girls collect during the summer immense quantities of the reindeer-moss from the fjelds; and when the autumn storms sweep the snow down the sides of the mountains, and cover up with its smooth uniform surface the steep and almost impassable roads, the farmer brings the moss, frozen into hard compact masses, on sledges down into the valley, where it forms an essential part of the winter fodder of the cattle in this district.

After a fatiguing drive of about three hours, exposed to the scorching sunshine on bare treeless moorland, we came down to a station hidden in a nook of the Nord fjord, called Bryggen. The whole of this region is beyond the ordinary tourist's ground, and is quite fresh and unexplored. Mr. Murray's guide has not penetrated into the scenery of the Nord fjord, some parts of which are truly grand and Alpine in character. Here we were admitted for the first and only time into the bosom of a Norwegian family. On all other occasions, travelling on frequented ground, we were treated as tourists, and got our meals in our own rooms. But here we were treated as guests and dined with the members of the household. If it was "pot-luck" we got, the proprietor must have been uncommonly well off to keep such a table, loaded with fish, flesh, and fowl. Our hostess did not sit with her husband and children. She brought in the dishes, and attended to the comfort of the guests. This created an unpleasant feeling in our minds; but apologies or entreaties to sit down with us would have been misplaced, as in Norway the lady of the house considers it her especial duty to superintend the operations of her servants, and make her guests perfectly comfortable.

In this part of the Nord fjord there were little creeks, where the shore sloped gradually down into the profounder depths. In this shallow water grew large quantities of wrack, dulse, tangle, and other common sea-weeds. Owing to the great depth of the water, into which the rocky shores descend abruptly, these sea-weeds are rare in Norway. Only in one other place did I notice anything like the sight which our weedy sea-shores present when the tide has ebbed. Usually there is but the slightest fringe of sea-vegetation marking the water-line along the rocky shore; and in many of the fjords even this is absent. At the head of the Sogne fjord, upwards of 120 miles from the open sea, there are no sea-weeds lining the precipitous shores. The water at the surface is almost fresh; indeed, I saw a sailor putting down a bucket into this stratum and drinking the contents. The influence of the tide is little felt; and the river that empties itself into it overlies the heavier salt water, and prevents by its intense coldness and freshness the growth even of the green ulvas and enteromorphas which in this country mark the junction between fresh and salt water. Owing to the absence of vegetation, fish and other fauna of the sea are rare; so that the inhabitants have not this source of supply to eke out the scanty produce of their miserable corn and potato fields. In the creeks of the Nord fjord, however, there was an unusual abundance of shell-fish and other forms of sea-life lurking among the dark tufts of fuci and tangle. I gathered a few specimens of *Natica Grœnlandica*, an Arctic and circumpolar mollusc, which becomes rarer and smaller towards the south; and of *Pecten islandicus*, which does not reach Britain. While gathering these northern shells, I thought of the remarkable parallelism between the distribution of the Arctic fauna and flora in Britain. Just as we have the remains of an Arctic flora, once overspreading the whole country, on the summits of our highest mountains, so we have the remains of an Arctic fauna which peopled all our seas during the Glacial epoch in the profoundest depths of our western sea-lochs, such as Loch Fyne and the Kyles of Skye. A little south of Tarbert, Loch Fyne deepens into a basin 624 feet below the surface of the water, a far greater depth than that of the sea outside, and clearly indicating that this narrow inlet is a submerged land valley, whose bed, if sufficiently upheaved, would be marked by a freshwater loch, like Loch Lomond. From this profound abyss Professor E. Forbes and Mr. McAndrew, in 1845, brought up with the dredge an extraordinary assemblage of molluscan animals, eminently Arctic in their character, once common in all our seas, ranging from the shore-line downwards. When the beds of these glacial seas were upheaved, several of the more delicate molluscs perished under the change of conditions, while others more accommodating survived. As the climate became more genial, the northern and Arctic shells that lived in the littoral zones retreated northwards, driven out by the migration of more temperate forms. Those that had greater capacities for vertical range, however, remained behind in the deepest parts of our sea-lochs, where the conditions of temperature were still suitable; and to this narrow range they are now restricted. The extreme scarcity of these Arctic shells in a living state, and the comparative abundance of dead valves, seem to indicate, as

Professor Forbes suggested, that the species thus isolated are now slowly dying out: so that the time may not be far distant when the last of the Arctic forms of the mountain-top and sea-bottom will disappear before the inroads of plants and animals of a milder climate, that will spread uniformly over all parts of land and sea. In connexion with the mollusca of Norway the singular fact may be mentioned, that some of its characteristic Arctic species are found as fossils in Italy and Sicily; and that other perfectly identical species are found living at the present day in the Mediterranean and Adriatic and in the North Sea, which are absent on the intervening coasts of the Atlantic, the only route by which, according to the present arrangement of Europe, they could have reached the one locality from the other. Among the living species common to Italy and Norway are *Nephrops Norvegicus*, *Lota abyssorum*, *Sebastes imperialis*, *Macrourus cœlorhynchus*, and two shells found by Professor Sars of Christiania, in the sea at Bergen, *Cerithium vulgatum*, and *Monodonta limbata*. The presence of these mollusca in the Mediterranean and in Norway, with their absence from the intermediate coast, is supposed to be owing to a connexion that existed during the Post-Pliocene period to the east of Europe, between the Mediterranean and the North Sea, which was interrupted at a later period by the elevation of the Alps.[1]

[1] This theory is still further confirmed by the flora of Sweden. Several of the most characteristic plants of Gothland, an island in

At nine o'clock at night a Government steamer employed in the postal service, and carrying on the traffic with all the stations on the Bergen route, appeared in sight. We rowed out to it in a small boat, and then steamed down the fjord, through the most intricate labyrinths of hills and islands. There is one rock rising 1,200 feet perpendicularly from the water, shaped like a huge cathedral with a gigantic tower at either end. It is called Hornelen, and our steamboat was named after it. It is the loftiest and most massive sea-cliff in Norway south of the Luffoden Isles. A great slice of it had fallen down two years previously, about two hours after a steamer had passed. The scar was still fresh on its side, and the *débris* formed a talus bank at the foot projecting into the sea. The depth of the water in this narrow channel is said to be very great, there being no soundings for two thousand feet. After spending some hours on deck, admiring the wild and ever-changing scenery, and watching the giving out empty and taking in full herring-barrels at the different stations at which we called, we retired to our berths and slept till about seven o'clock in the morning, when we found ourselves among the

the Gulf of Bothnia—such as *Helianthemum fumana, Inula ensifolia,* and *Serapias rubra*—are identical with those of the limestone mountains of Austria; while the vegetation of the neighbouring island of Oland is of a decidedly Mediterranean, or even African, type. Among its rarer plants may be mentioned *Helianthemum Œlandicum, Carex obtusata, Artemisia laciniata, Anemone sylvestris, Ulmus effusa,* and *Viola persicifolia.*

skerries on the coast, within forty miles of Bergen. These rocky islets are very remarkable. They occur in countless numbers all along the coast from Christiania to the North Cape, and though composed of gneiss afford a striking proof of the tremendous abrading action of one of the stormiest seas in the world. They are of various sizes, from a huge boulder barely rising above the level of the water, to lofty castellated crags many acres in extent, and are either bare or covered with shrubs or fir-trees. Between them the sea winds in and out in the most intricate fashion, and they are so like each other that it is astonishing how the pilot can thread his way among them. There is never any of that sloping which distinguishes the shores of other countries. Quite close to the rocks the depth is in some places unfathomable. In many instances the narrow creeks and channels run far inland, so that it is frequently necessary to journey a hundred miles by land between two places not more than two or three miles apart in a straight line. Many of the skerries are shaped like the Devonshire tors; they are what are called in geological language *roches moutonnées*, rounded, smoothed, and polished hummocks, moulded by the passage of a thick body of ice over them during the Glacial epoch, and marked, many of them very distinctly, by close parallel flutings, indicating the direction of the moving ice. From these glacial markings, where no ice is now to be seen, we can trace by the characteristic evidence of striæ, moraines and boulders, the course of ancient glaciers up to the great ice-fields of Justedal and the Folgefond, still existing in the interior. We must regard the present glaciers of Norway as the shrunken remains and silent witnesses, in a milder climate, of immense glaciers which at one time stretched down and filled each valley, and went out to sea like the glaciers of Greenland at the present day.

The glaciated skerries, judging from the profound depths of water around them, are the tops of submerged mountains, and the fjords that wind among them deep glens that have not yet fairly risen out of the sea. That Norway has been slowly rising from the sea within comparatively recent times is proved by many indisputable signs. On the shoal or bank which lies out in the Christiania fjord to the west of Drobak, and which is from sixty to ninety feet deep, there are immense masses of a peculiar coral called *Oculina prolifera*, firmly attached to the solid rock, though dead and stripped bare of its formative polyps. This coral is found on the western and northern coast of Norway in a living state, only at the vast depth of from 1,000 to 2,000 feet, where it forms large, bush-shaped clusters about two feet in diameter. The fact of its occurrence in a dead state on the Drobak bank proves beyond doubt that that bank was elevated to the extent of at least 800 feet, when the polyps, incapable of bearing the increased temperature of the shallower water, died *in situ*. On the same bank, also in a dead state, is found the *Lima excavata*, a species of shell-fish which lives only in the region of the deep-sea corals, at from 150 to 300 fathoms. Professor Forbes and Mr. Robert Chambers speak of "the great freshness of the raised terraces which stretch at various heights along the coast, as if to show where the surf had beat during prolonged intervals in the course of upheaval." On these terraces vast quantities of shells are frequently found identical with those living in the neighbouring seas, and looking as fresh as if they had been cast ashore only yesterday. Brogniart found balanus shells on the solid rock at Udevalla, on the Swedish coast of the Cattegat, 200 feet above the present level of the sea, and Keilhau near Hellesda, in Aremark, 450 feet above the sea. The last accomplished geologist pointed out to Mr. Robert Chambers serpulæ still adhering to the face of a rock about a mile from Christiania, 186 feet above the surface of the fjord. We have thus the best possible proof of an elevation of the land during the existence of its present fauna.

Every traveller is greatly struck with the resemblance, only on a larger scale, between the coast of Norway and the coast-scenery of the West Highlands of Scotland. The same causes, acting in similar circumstances, produced this resemblance. In Scotland these causes have long been quiescent, and we can only speculate and theorize regarding their mode of action in the remote past. In Norway they are still in operation, and their modifying effects may be seen fresh and recent in many places. Norway may be regarded as a connecting link between the present state of Greenland and the state of Scotland during the Glacial epoch. When Scotland had its glaciers and snow-fields, Norway was completely enveloped in ice; and now that the line of perpetual snow has gone beyond the summits of our highest hills, we recall in the perpetual snow regions of Norway the appearance of our own country at the close of the Glacial epoch, when the glaciers were retreating from the coast into the high grounds of the interior. Not only in geological development, but also so far as progress during the historical epoch is concerned, Norway may be regarded as "a larger Scotland post-dated," a country still in its green youth, while Scotland is in its old age. The forests that overspread its surface at the present day are like the extensive forests of Scotland during the Roman invasion, whose remains are found in our numerous peat-mosses. The existing Norwegian fauna once roamed in our woods and hills. In Caithness the reindeer lingered until about the beginning of the thirteenth century. The Orkneyinga Saga relates that the Jarls of Orkney crossed over to the mainland to hunt it in the twelfth century. Acccording to tradition, the last wolf in Scotland was slain in 1680 by the famous Sir Ewan Cameron of Lochiel. There is ample evidence to prove that the brown bear lived in this country less than a thousand years ago. Up to the middle of last century the capercailzie, or great cock of the woods, the largest member of the grouse family, abounded in our woods. It disappeared with the destruction of the Caledonian forest, the cones of which formed its principal food ; and though it has been reintroduced from Norway, it is confined to one or two districts, where it is almost as tame as a barn-door fowl. In the highest solitudes of the Grampians still linger the Alpine hare and ptarmigan, the last survivors of the ancient Norwegian fauna of our country, which owe their preservation to their power of adapting themselves to their circumstances, changing the colour of their fur and plumage, a provision which not only regulates the temperature of their bodies according to the changes of the seasons, but by assimilating them to the prevailing colours of the scenes amid which they live, enables them to elude the keen eyes of their numerous enemies. Thus the wild animals of Norway are those which formerly lived in Scotland, but are now nearly all extinct. The manners and customs of the Norwegians in the remoter districts are also those of our ancestors several hundred years ago, and their udal system of land proprietorship is that which existed in Scotland prior to the introduction of clanship and feudalism, and the remains of which may still be seen in the existence among us of "bonnet lairds," similar to the Norwegian "bonder," who cultivate the small properties which they inherit. Thus a visit to Norway gives the Scotchman an admirable idea of the appearance of his country and the condition of his ancestors in the Middle Ages.

No incident of any moment occurred while we threaded our way in and out among the skerries—except that at breakfast in the cabin a member of our party who had a peculiar *penchant* for sausages persuaded one of the ladies to taste a particular kind, on the plea that he had found it exceedingly nice. He asked the waiter if it was made of reindeer venison, and grew somewhat solemn—as well as his fair Eve—when told that it was genuine horse-flesh! We arrived at Bergen at two o'clock, and were delighted with its picturesque appearance and romantic situation. It is built upon two bays of the fjord, with a narrow point of elevated land between them, on which stands the fortress of Bergenhuus, where formerly stood the palace of King Olaf, the founder of the town. It is surrounded by steep and rugged mountains, between two and three thousand feet high, so that you have all the bustle of a commercial town quite close to the loneliness and grandeur of an Alpine solitude. The narrow harbour is always crowded with shipping, and the suburbs at the base of the mountains are occupied with gardens, and country villas embosomed among woods, with green lawns sloping down to the fjord. Outside the town there is a magnificent avenue of old linden trees about a mile long, from whence beautiful glimpses of the surrounding scenery may be obtained. This is the most northern limit of this tree, and yet it is as full grown and majestic here as in the avenues of England or Germany. We heard a sermon in the fine old cathedral, and inspected the antiquities and objects of natural history in the museum. The antiquities are principally sepulchral urns, arms, Runic inscriptions, Norwegian coins dating from the time of Haco the Good in the tenth century, and a curious old Byzantine picture presented to one of the churches in the Sogne fjord in the eleventh century by a sea-king who had procured it from Constantinople. I observed that there were no contributions to this department from the Arctic provinces. Worsa, the director of the museum of northern antiquities at Copenhagen, informed me, while conducting me over the rooms devoted to relics of the Stone period, that all these stone implements came from Denmark and the southern parts of Sweden and Norway—none having been found in the northern parts. The inference is therefore clear that these northern provinces were unoccupied by man during the earliest ages of British and continental history. This theory coincides with the evidences of physical geography, and points to the gradual amelioration of the climate in these regions.

We were greatly amused by the extraordinary character and variety of the costumes of the peasants from the surrounding districts, who came in to Bergen to do their marketing. Some of these peasants are said to be of Scotch extraction—a large colony of Scotchmen having settled about the twelfth century in the neighbourhood of Bergen. Some of the women had white shirt-sleeves, scarlet jackets, gorgeous breastplates of coloured beads, and white caps of the most extraordinary shapes and dimensions. Others had green jackets and dark caps. It is a great pity that both in Switzerland and Norway the picturesque costumes of the peasants should to so large an extent be abandoned for the uniform and unmeaning dress of all classes throughout Europe. We paid a visit to the fish-market, which is one of the most interesting sights of the place. All the fish—of which there is an immense variety—are brought in alive, and kept swimming about in tubs of salt water until purchased; for a Norwegian would never think of buying a dead fish; he likes to be assured by more senses than one that it is quite fresh. Among the fish we noticed grey gurnards, *torskwrasse*, of many colours, and coalfish (*Gadus carbonarius*) in shape and size like a salmon, with a black back and a silvery belly. There were also a few specimens of the bergelt, or Norwegian haddock (*Sebastes Norvegicus*), somewhat like a perch, which exhibits all the hues of the gold fish. It is caught in very deep water with long sea-lines, and is considered a great delicacy. Bergen has the reputation of being one of the rainiest places in the world. The average number of rainy days in the year is said to be 200. This extreme humidity is shown not only in the actual amount of the rainfall, but in the almost constant presence of large quantities of aqueous vapour in the atmosphere, even on days that are considered clear and bright. This constant wet blanket spread over the town, if it does damp the joys of the inhabitants and engender melancholy "vapours," gives compensation by greatly modifying the severity of the winter. Probably the range of temperature throughout the year is smaller and the mean annual temperature higher in Bergen than in any other place so far from the equator. During our residence, however, there was not a cloud in the sky, which was as deep-blue and transparent as that of France or Italy. The heat was tropical, and we had to dodge the sun continually in our walks through the dusty and glowing streets. The mere effort of writing a letter in a room where all the windows were thrown wide open threw me into a profuse perspiration. I love to look back upon the wonderful beauty of the nights we spent at Bergen. From my bedroom window I looked out for hours with intense enjoyment of the scene. Below, the gaily-painted houses looked ghostly in the tender twilight that brooded over them; above, the moon shone large and golden in the blue languid sky, casting down a path of light along the surface of the placid fjord. The mountains, mellowed down and empurpled by the sunset, reposed with a dream-like beauty on the near horizon; while the stillness was broken by sounds that harmonized with it—the ripple of a passing oar, or a simple song heard clear and distinct from afar. The whole air and appearance of the place at such a time reminded me more of Oriental cities described in the Arabian Tales, than a matter-of-fact Norwegian town, crammed with odoriferous stock-fish and casks of cod-liver oil.

CHAPTER V.

THE SKJEGGEDAL-FOSS IN NORWAY.

SOME one has remarked that mountains present peculiar attractions to men, while women find in waterfalls something more congenial to their nature. This, like a great many other general statements, is probably too large an induction from the facts of the case. It is true that the membership of the "Alpine Club" has been confined exclusively to the male sex; but, in districts favoured with a famous waterfall, it has not been found that the fair sex have monopolized the services of the local guide. Ever since the discovery of the picturesque in nature, during the present century, both sexes seem to have shared indiscriminately in the admiration which mountains and waterfalls call forth. If there be any preference shown by women, it is perhaps due to the fact that waterfalls are more accessible than mountains, and do not require for their cultivation specialties of dress and muscular development. Notwithstanding this, however, it seems to me that there is a measure of truth in the aphorism. I believe that the mountain does harmonize more with the masculine than with the feminine character. Its ruggedness, solidity, height, and changelessness symbolize peculiarly manly qualities; while the toil, patience, and endurance needed in its ascent are exercises in which man delights. It appeals in its form and associations to his sense of power and self-reliance. The waterfall, on the other hand, speaks more to the gentleness and softness of the feminine nature. It is moulded by the form of the rocks and by the play of the winds, and it yields itself gracefully to the influences of its circumstances. The continuous murmur and fall of the snow-white water; the unity and variety of the forms which it presents; the quick play of light and shade on its surface; the rainbow that opens its blossom of light amid its spray; the tender and graceful vegetation which its perpetual moisture nourishes around it, from the aspen that trembles to its shout, and the birch that hangs its tresses in its foam, to the moss that cushions the ledges of its rocks, and the lichen that makes its cliffs hoary: all these features of the waterfall appeal to qualities that are more often found in woman than in man.

Waterfalls, however, are very varied. Some are quite as masculine in their character as mountains. They have no soft and graceful surroundings. They do not hide themselves in the loneliest recesses of glens, protected by cliffs and shaded by foliage; but leap in the open light of day, in straight lines, from the brink of naked precipices.

Grandeur and sublimity are their sole characteristics. In proportion to their height and volume, they lose their picturesqueness and beauty; and appeal to another order of feelings in the human bosom. This is especially true of the Norwegian waterfalls. They possess grandeur, but not beauty. Owing to the peculiar formation of the country, it is rare to find that gradual sloping of the stream, and that succession of leaps, fringed with trees and shrubs, which contribute so much to the picturesqueness of a waterfall. The mountains are immense tablelands or plateaus, terminating abruptly on both sides in lofty mural precipices. Consequently the streams that are formed on them from the melting of the snows in summer, after running a short course, fall sheer down into the glen or the fjord; whereas in this country or in Switzerland the mountains are constructed, not in the embattled style, but on the ridge and furrow principle, and slope gently into the valleys, so that the streams that gather in their bosom flow gradually down, increase as they flow, and form a succession of waterfalls, according as they meet with rocks in their course. The waterfalls of Norway are thus necessarily higher than those of any other part of Europe; but they want the fringing of woods and the concealment of picturesque rocks peculiar to more gradual falls.

Waterfalls are also more numerous in Norway than they are anywhere else. "The mountains," to use the expressive language of a Belgian tourist whom I met at Utne, "are peopled with them." They lend animation to every scene, and hang from every cliff their scarf of liquid drapery. Hundreds of cascades unknown to fame, though far higher and grander than the Marboré fall, near the Gavarnie, in the Pyrenees, or the too celebrated Staubbach in Switzerland, may be seen in the course of a single day's journey in the interior. The Riukan-foss, or Reeking Fall, in Upper Thelemarken, drops almost perpendicularly about 800 feet into a gulf so filled with vapour that its bottom cannot be seen. The body of water is very considerable, being the overflowing of the Miöswasser, a lake thirty miles long and more than two miles broad. The Sarpen-foss is grander than the falls of Schaffhausen on the Rhine, being formed by the united waters of the Lougen and the Glommen, the two largest rivers in Norway, which drain the whole of the east side of the country for more than 300 miles. The height of the fall is eighty feet, and almost equals in volume of water the famous Trollhatta Fall, by which Lake Venner in Sweden empties itself through the Gotha-Elv into the Cattegat. The most numerous as well as the finest waterfalls in Norway, however, are to be seen in the Hardanger district. In this region are the Rembiedals-foss and the Skyttie-foss—both very magnificent falls, though situated in remote out-of-the-way glens, and therefore visited by few travellers. Here, too, is the better-known Ostud-foss, which falls into the depths of the Steindal valley, not far from the station-house of Vikör.[1] But by far the most celebrated of the waterfalls of the Hardanger is the Vöring-foss, said indeed to be the grandest cataract in Europe, and the lion of Norway. Its height is upwards of 900 feet, and its volume of water fully larger than that of the Handek in Switzerland. A Frenchman on one occasion was so excited at the thought of visiting it, that even when his steamer entered the Hardanger fjord, nearly a hundred miles distant, he broke out in a transport of enthusiasm, "I am coming near it; I am coming near; for thirty year I dream of Vöring-foss." The spectacle is indeed grand beyond description; but it labours under the great disadvantage that it cannot be seen from below. I believe that one or two daring cragsmen succeeded in getting pretty near the foot of it; but their view of the waterfall was greatly obstructed by a projecting rock. The ordinary tourist sees it from the edge of a great precipice at a considerable height above the top of the fall. Keeping a firm hold of the guide's hand—if you have sufficient nerve and are not oppressed with giddiness—you can bend your body half over, and look down into the awful abyss filled with seething waters and blinding mists. A vision of a great white mass of foam falling, minute after minute,

---

[1] Murray has evidently taken his description of it on trust from some imaginative native, for the proportions which he gives are vastly exaggerated. It pours over its rock certainly more water than "a gill in a minute," but, like Southey's "Force of Lodore," it is very disappointing to the eager visitor.

pausing as it were at intervals in mid-air, but still falling down, down, far out of sight into the bowels of the earth, with a roar that seems to shake the rocks to their foundations, is caught during the frenzied gaze and photographed upon the memory for ever. Woe betide the unhappy tourist who is seized by nightmare the first time he goes to sleep after having stood on this giddy height!

Within the last few years a waterfall has been known to the tourist-world which promises to rival the Vöring-foss. The Skjeggedal-foss—for such is its jaw-breaking name—is not nearly so high; but the body of water is larger, the scenery is more savage, and it can be approached quite close and seen in all its grandeur from the foot. Opinions are very much divided regarding the claim of each to pre-eminence. Very few have visited the Skjeggedal-foss; and therefore the notices of it are exceedingly scanty. Murray does not mention it at all. In the "Dag-boks" at Vossevangen and Odde I found it praised by one in the most extravagant terms as decidedly the finest waterfall in Norway, while another entry was to this effect: "The Skjeggedal-foss should be seen before and not after the Vöring-foss."

Happening to be with a party of friends at Odde, a small picturesque village at the extremity of the Sor fjord—a branch of the Hardanger, and the nearest starting-point for the Skjeggedal-foss—I determined to judge for myself regarding the merit of this cascade. Accordingly, accompanied by a friend, I set out on Wednesday the 17th July, at seven o'clock in the morning. We secured the services of Lars Olsen, a native of Odde, who discharged the duties of guide throughout the day in the most admirable manner, and whom we have therefore much pleasure in recommending to future travellers. There was all the interest and excitement of discovery about our adventure. It was something to boast of, to be able to visit a scene unknown to the ubiquitous Murray, to act as the pioneers of that paladin of modern times, and perhaps add an interesting paragraph to future editions of the well-known red Handbook of Scandinavia. The morning was all that could be desired. A few clouds threatened at first to discharge their watery burdens, but they soon passed off, and the sun shone brightly in a blue and unclouded sky. We laid in a comfortable stock of provisions, as the excursion, we were told, would occupy the whole day. Two smart fellows from Scotland recorded the fact that they had done it in eight hours, but sensible men who were not walking for a wager, and who preferred enjoying scenery to doing it under steam-pressure, gave their evidence that it could not be managed in less than twelve or fourteen. The latter verdict we found from our own experience to be the true one. We brought with us waterproofs on account of the lowering appearance of the sky at starting, but we found them very serviceable afterwards, enabling us to approach nearer the waterfall than we could otherwise have done, without being drenched by the spray. Lars had his provender carefully rolled up in a coloured pocket-handkerchief; it consisted of about six square feet of *fladbrod*—a kind of very thin barley-scone—and a small piece of raw mutton dried into the hardness and colour of a mahogany slab, and needing no further cooking.

Stepping at the quay into one of those rickety Norwegian boats, sharp at both ends, which are so alarming at first to timid sailors, we rowed up the fjord for about four miles. The sea here is very narrow, and the banks on both sides very steep and lofty. At the foot of the left bank are green patches of cultivated land here and there, and clusters of picturesque red wooden houses; in the higher region pines and birches fringe the ledges of the rocks; while on the sky-line the great glacier of the Folgefond shows its white teeth in every hollow between the cliffs. In some places the glacier was suspended over the edge of a precipitous rock far up in the air, and one felt afraid in passing underneath lest the huge mass should be loosened and fall down with a mighty plunge into the fjord. Many of the houses look as if they lay directly in the path of the avalanches, great talus-heaps of *débris* lying perilously close to them. The overhanging tongues of ice were very beautiful, being much crevassed, and showing in every wrinkle and hollow that marvellously vivid sapphire colour with which every glacier-student is familiar. Nothing could exceed the purity of the ice, or the stainless whiteness of the snow—in this respect presenting a striking contrast to the discoloured glaciers of Switzerland, whose dirty faces no amount of Alpine rain can wash clean. Some years ago, when the supply of ice in London was nearly exhausted, a ship was chartered to the Hardanger, and brought home a cargo of magnificent fragments of the Folgefond glacier. Though the experiment answered admirably in every way, I am not aware that it has been repeated.

Calm and still as the morning was, we did not hear the tinkle of the bells of the lost seven parishes said to be buried on account of their great wickedness under the everlasting snows of the Folgefond, and which many superstitious ears have heard on certain propitious days. This tradition is very similar to that of the Blümlis Alp in Switzerland, and, like it, is evidently not altogether a myth. It tells of a change of climate, and of a gradual advancement of glaciers, overwhelming districts once fertile and inhabited, of which many traces may be seen in the physical appearances around. For hundreds of years the Folgefond glacier is said to have remained stationary, but it is most certainly advancing in one direction; for during a visit to the small tongue which descends on the east side to within a thousand feet of the sea-level, called the Buerbræ iis, I saw the ice distinctly moving, stones falling from its edge, and the ground newly ploughed up before it. The right bank of the Sor fjord is more precipitous than the left, though not so wild and Alpine-looking. Huge masses of broken rocks are piled above each other, like a Titanic battle-field, at the edge of the water. Bright green birches, with uncommonly white stems, are interspersed among them, and soften their harshness, while high overhead the precipices form a gigantic wall, with a fringe of pine-trees gleaming along their ledges in the sunlight, like the spears of a celestial army. Little streamlets on both sides flow down the rocky gullies in one long continuous line of foam from the clouds to the sea, and make a pleasant all-pervading murmur in the air. The water of the Hardanger fjord in this place is of a deep green tint, and in the chart is marked as upwards of a thousand feet deep. There is no shelving shore, but the rocks go straight down into the profound depths.

After two hours' rowing through this magnificent scenery, we came, on the right bank of the fjord, to the entrance of a wild gorge, through which flowed the foaming waters of the Skjeggedal torrent. An enormous wall of rock rose up on the left side without a ledge or a break, destitute of the slightest tinge of verdure. On the other side the precipice was more sloping, and admitted here and there of a few clumps of birches and pines growing on its shelving sides. The mouth of the gorge was filled with great banks of *débris* brought down by the stream in the course of ages; and on these, which were carefully cultivated, stood a small but very neat-looking hamlet, called Tyssedal. The people were busy hay-making — gathering the natural grass, and piling it, to dry in the sun, on the upright framework of wood which is erected as a permanency in every hay-field in Norway. Two or three sunburnt girls, with green bodices, white sleeves, and unusually large picturesque-looking caps, were singing a wild Norwegian *jodl*, while tossing about the hay. The position of this hamlet struck us as exceedingly precarious. It seemed to fill up all the available space in the gorge, and it looked as if a storm of more than ordinary severity might have washed both houses and fields down into the sea.

Crossing the foaming torrent, which for half a mile agitated the placid waters of the fjord, we moored our boat in a sheltered creek, and stepped on shore. As we entered the frowning portals of the gorge, leading to the great inner mystery of the waterfall, we had a feeling of strange awe, such as the Assyrians of old must have experienced when passing between the monstrous human-headed bulls that guarded the gates of their temples. Our course at first lay up the steep bank of the river on the right hand, through a fine wood of Scotch firs, whose great red trunks and rich green foliage would have done credit to any nobleman's park. The sun shone through the openings between the trees in bright belts of gold on the mossy sward, crowded with myriads of whortleberries, whose glossy leaves and clusters of white bells excited our admiration. I never saw such a quantity of the beautiful *Linnæa borealis*—named after the immortal botanist, and called by the Norwegians *windgräs* — growing anywhere as in this wood. Its modest pink blossoms covered every available space, and its rich fragrance pervaded all the air, producing, along with the resinous scent of the firs, a peculiarly delightful and exhilarating impression. Fine specimens of the *Melampyrum sylvaticum*, with flowers larger and yellower than those of the same species in this country, bloomed on every side. Ferns abounded: clusters of the tall *Struthiopteris germanica*, lovely patches of the fragile oak fern, and, above all, large tufts of the *Woodsia* growing everywhere among the stones. This last fern, which in this country is only found in two or three remote localities among the loftiest mountains, is very common and abundant by the waysides in many parts of Norway—indeed, as common almost as the *Polypodium vulgare* with us. There were also numerous anthills, formed of the dry needles of the fir, like those with which the tourist is familiar in the pine-woods of Braemar. Some of these were of enormous dimensions, and their tenants were swarming in myriads on the outside, running up and down to warm themselves in the sunshine. A stick thrust into one of the hills smelt overpoweringly of formic acid.

Gradually, as we passed through this enchanting wood, where everything was left to fall or grow in the wild yet graceful disorder of nature, the path became steeper and less defined. In some places it consisted only of a tree-trunk fixed along the sloping side of a granite rock by an iron bolt. Over this precarious footway we practised successfully a series of tight-rope performances. We were struck with the curious appearance of some of the nearer gneissic rocks, forming bands of thin, regular strata, lying over each other, exactly like the huge, unshapely slates on the roofs of Norwegian houses, or the armour-plate of a man-of-war, and covered with the black stains of a species of *Lecidea*. On emerging from the wood, we found ourselves on a kind of plateau of bare rock, without a particle of vegetation—not even a lichen or a moss to tint its surface. It was perfectly smooth, and sloped rapidly down at a perilous inclination for a few yards, terminating abruptly in a precipice. Across this steep slope the guide walked without a moment's hesitation, his flat shoes catching firm hold of any roughness in the rock. I followed mechanically, though not without considerable trepidation, for the soles of my boots were very thick and slippery, and I knew that if I lost my footing I could not recover it, but would be hurled with fearful momentum down the slope into the abyss. One shuddering glimpse I caught of a wild whirlpool of waters far below made my blood run cold; and as in this case discretion was the better part of valour, I am not ashamed to own that I willingly submitted to "a spirit of infirmity," and crawled on all fours. To make matters still worse, we had to ascend, about the middle of the passage, to a higher stratum of sloping rock by means of a fir-trunk, with notches cut in the side of it for steps. I need hardly say that I breathed more freely and saw more grandeur in the scenery when we reached the other side of this dangerous roof. The pathway after this was along the edge of a precipice, but its terrors were concealed by a profusion of trees and bushes. In a wider space, we came upon a man and his wife busy erecting a wooden hut from the materials on the spot. An axe was their only tool, and it was wonderful what a shapely framework they had constructed by its means, without any nails. We asked them what induced them to build a house in such a spot. It could not be a saeter or hill-farm, for there was no grass around, and no possibility of housing or feeding cattle on such a precipitous slope. The man replied that it was intended to be an inn—I suppose the "Hôtel du Skjeggedal-foss." It seemed a very hopeless speculation in present circumstances, but it was an idea worthy of the genius that first thought of an inn on the top of Snowdon, on the Riffelhorn, or the St. Theodule Pass, and deserved from its very boldness and originality to succeed. Perhaps this sketch may be the means of bringing custom to the place. If so, the only commission I shall expect as a touter from the hotel-keeper and his lady is a bottle of the best Baiersk Ol the next time I pass by their hospitable door, for in my thirst and fatigue I grievously missed it on this occasion.

We had now reached the highest point of the

ascent, and were congratulating ourselves that all danger and cause of fear to weak nerves were past, when we came to a staircase that beat all structures of the kind I have ever seen. It descended for about twelve yards at an angle of some fifty-five degrees, and consisted of rough irregular steps projecting an inch or two beyond each other. On the one side was a lofty wall of rock dripping wet, and covered with bright green mosses and gelatinous masses of vegetable growth, so that there was very little hold for the hands, while on the other there was a sheer precipice, and far below a raging torrent falling into a hideously black linn; and from this danger there was nothing, not even the slightest handrail, to give one a feeling of security. It was a place to try the nerves even of a member of the Alpine Club. We crawled down, clinging to every projection with tooth and nail, the calves of our legs all the time trembling like a jelly. When we got safely to the bottom, we thought that we had accomplished a feat to be proud of all our days, but our vanity received a severe shock when the guide, looking back upon the staircase, said in the most matter-of-fact voice, "*Det er ond plads for hesten*" (That is a bad place for horses). After all, we had only done what a quadruped was in the habit of doing; though how a great long creature like a horse could manage to come down this breakneck place, with nothing to cling to, was a puzzle which I cannot yet understand. I can only say that I should like to see him at it. Astley might get a new idea from it. There is a kind of saeter, or hill-farm, farther up the gorge; and its dairy produce, it seems, is strapped on a horse, and thus carried down to Odde, where it is sold for groceries and other needful articles, which are brought back in the same picturesque fashion. Of course, no one could ride on horseback along the path by which we had come, although we found an entry in the "dag-bok" at Odde, complaining bitterly that the innkeeper had refused to give horses for the excursion to a lady and her husband! We had previously seen in our carriole-travelling some of the remarkable feats of the Norwegian pony, but we had no idea he was capable of such an Alpine Club exploit as the descent of this staircase, and we vowed a vow on the spot that nothing would ever induce us to venture upon a path which a Norwegian pony could not traverse—a vow which we religiously kept. We had now got over two very bad places, but of course we had to go back, and the thought of returning in the same way did not add much to our peace of mind or enjoyment of the scenery. It was the sword of Damocles suspended over our head.

The descent from this staircase was very rapid, but it was over very rugged ground. We picked our way in and out among chaotic masses of large loose stones, placed at every possible angle, but generally the sharpest side uppermost. At last we came unexpectedly upon a little oasis in the wilderness—a quiet nook of cultivated ground. The space here was wider, the rocks having retired to a greater distance, and allowed more of the blue sky to be seen, and the sun to shine down unobstructedly in all his warmth and golden splendour. This miracle of refreshing greenness and beauty was evidently the slowly-accumulated deposit of the denuding power of the stream. The soil, though light and shallow, yielded a fair crop of potatoes, and the grassy pastures were golden with buttercups, and sprinkled with white honey-sweet clover blossoms. A cluster of rude wooden houses stood on the spot amid clumps of graceful birches. A little tarn stretched out in front, into the head of which tumbled down an enormous body of foaming water from a considerable height, while the other end of it, a little way down, discharged a powerful torrent that had to force its way through a very narrow passage in the rocks. In the struggle, the water presented a most lovely appearance, broken up and churned into snow-white billows tinged with the brightest cerulean hues, like the interior of glacier crevasses. It was a sight that had a terrible fascination about it, and from which it was most difficult to withdraw the eye. As we were gazing, spell-bound, a strange specimen of humanity came up to us with a peculiar duck-like waddle. He was a young man apparently about eighteen years of age, blind, dumb, and idiotic. He had no chin, and his face had the strange bird-like look which we see in the hieroglyphic paintings of the Aztecs, or in South American antiquities. He was conscious of the presence of strangers, but he had no sense to which we could appeal, and we were therefore compelled to pity his wretched condition in silence.

The house into which we entered was that of the *bonder*, or peasant proprietor, and was far superior to the others. The whole gorge of the Skjeggedal, eight miles in length, and I know not how many in breadth, belongs to this man as "udal-land," paying no acknowledgment, real or nominal, as a feu duty or reddendo, possessed consequently without charter, and subject to none of the burdens and casualties affecting land held by feudal tenure. But as this property consists principally of rock and water, it is not very productive. There is a great supply of timber, however, and the quantity annually cut down and floated on the river to the Hardanger ought to yield him a comfortable income. He informed us that he had nine milch cows, three horses, and twenty sheep, all finding a precarious subsistence on the grassy patches laid like green carpets on the sloping *débris* of the rocks. He had under him three or four married farm servants, holding cottages beside his own with a small portion of land, rent free, but under the obligation of working for him during a certain number of days in the year. Our "bonnet laird" had a wife and family of four small children, as shy as the ryper or ptarmigan of the fjelds. They were very unlike the inhabitants of a civilized world in look and dress, and so unaccustomed to visitors that on our appearance they fled and hid themselves behind the maternal wing. The gudewife—a very slatternly woman, with a patient, depressed face—offered us a drink of rich milk. The room was large, but very bare and cold. Its only furniture consisted of a curious cooking-stove, with Pompeian figures moulded in its iron sides, two rough bedsteads covered with reindeer skins, and a dairy press well filled with cheeses, butter, and bowls of milk. On the bed was a strange wooden dish, grotesquely carved, and painted in red, blue, and yellow, filled with a dark, muddy-looking liquor. This was a species of ale, prepared, instead of hops, with the leaves of a kind of ranunculus called *peast*, growing in miry spots. It is said to possess very peculiar intoxicating qualities, inspiring those who drink it with extraordinary activity and contempt of danger, but leaving a reaction of profound lassitude and debility. Tradition points to this beverage as that used by the famous Berserkir to inspire them with fury when going on their marauding expeditions. Our friend the farmer took a hearty draught of it, and offered it to Lars, who very modestly touched it with his lips, after having first shaken hands with his host and hostess, as the manner of the Norwegians is when receiving any favour. It was offered to us hesitatingly, but we shook our heads. It looked such a disgusting mess, that nothing could induce us to try it; and Lars assured us afterwards that it was as abominable to the taste as to the sight. We pitied the lot of these poor people, shut up in this wild dungeon among the rocks, far from their fellow-creatures, and isolated from all the refining and ennobling influences of civilization. The contrast between their winter and summer life must be very great. In summer their occupations are exceedingly varied, owing to the absence of all division of labour; and these are not shortened in this latitude by any interval of darkness; consequently they have that recreation in change of labour, which is perhaps the greatest enjoyment of a working man. But to this excessively active period succeeds a long winter of nearly nine months, during most of which there are only a few hours of daylight, while the frequent storms, and paths made impassable by snow and ice, prevent all communication with their nearest neighbours for weeks together. At such times their sufferings from enforced idleness and *ennui* must be very great. Indeed it is astonishing, considering the wild and gloomy character of the scenery, and the loneliness and monotony of their lives, that cases such as those of the wretched young man we met are not even more frequent. Scotchmen or Englishmen compelled to live in like circumstances would infallibly go mad; but the Norwegians are very patient and much-enduring, their tastes are simple, their wants few, and they have never known any other mode of life, so that custom reconciles them to what to us would be unendurable.

At this stage Lars had to resign his office: for the duty of conducting us to the waterfall now devolved upon the bonder. Going before us, therefore, we followed him past the hamlet, through fields purple with bluebells and the largest and loveliest pansies, over a rough wooden bridge, under which thundered a foaming torrent, forming a fine waterfall among the rocks high on the left. Dressed in knee-breeches of well-worn reindeer-skin, we greatly admired the symmetry of his calves, and the firmness and precision of his tread. His were the very legs of a mountaineer, accustomed to footing it in the most precarious places. A row of large silver buttons—made out of old coins, with the image and superscription of Frederick of Denmark still upon them—adorned his blue woollen coat, so that he was change for two or three specie dollars at any time. He brought us to the boulder-strewn edge of the tarn, and, launching his boat, speedily ferried us across the troubled waters. We landed on a plot of peaty ground, covered with tufts of beautiful cross-leaved heather in full rosy bloom, and white with the large flowers of the *Moltibôer*, or cloud-berry, which would afford many a delicious feast when the rich orange fruit was ripe. Clambering up by the side of a craggy knoll, over which the aforesaid waterfall precipitated itself,—so smooth and transparent at the top, before it was churned into foam, that the rock underneath could be plainly seen,—we came to the edge of another lake, four miles in length, and about half a mile wide, called the Ringedal's Vand. It is upwards of a thousand feet above the level of the Hardanger fjord, and is surrounded on every side, except where it discharges itself in the cataract, by lofty rocks which rise perpendicularly from the water's edge to a height of between two and three thousand feet. In a few places only is there any sloping ground formed of the *débris* brought down by waterfalls on either side; and such ground, covered with dwarf birches and bright green grass, formed a refreshing contrast to the dark frown of the barren rocks. I always looked out for such places, and had a feeling of relief when nearing them, for there at least I knew that I could scramble out and find a footing if anything happened to the rickety boat. Wherever there are any ledges or crevices in the precipices, there the hardy spruce and Scotch fir flourish. Hundreds of trees, with astonishing pertinacity, cling to the most fearful places, where there is hardly a particle of soil to nourish them; and their gnarled roots, fully exposed, crawling over the bare rock, look like the talons of a bird of prey. When passing by, close to the shore, we saw the farmer's servants perched above us on a precipitous rock, cutting down a huge fir, or lopping off its branches, and squaring its trunk for the market—their boat lying moored close by; while, on a projecting crag over the cataract, others of them were pushing with a long pole into the current the logs that had got jammed together in the back water. Both occupations looked very perilous, but the men seemed cool, smoking their pipes, and hailing us with a very cheery " gud-dag." Lars and the farmer took an oar each, and rowed us across the current to the other side of the lake in alarming proximity to the edge of the waterfall. None but strong and practised boatmen could hold their own here, and keep the boat in the right place. The breaking of an oar would be fatal. The water was cold as ice, and very deep, between one and two hundred fathoms, the bonder assured us. Its colour was dark indigo blue, the colour of the ocean when deepest; but in one or two places, where the depth decreased near a projecting promontory of boulders, it was of a rich green. Nothing could be more soft and tender than the gradations of this tint made by the water shoaling to the edge; gleams of malachite and emerald vanishing in transparent aqua-marine, and strangely interspersed with cobalt hues from the darker depths. It was a miracle of colour such as would have astonished and delighted a painter's heart.

Several waterfalls poured down the cliffs on either side, the finest of which was the Tyssestrengene. It was very peculiar, consisting of two distinct falls, formed by two torrents—separate, and yet blending strangely together. The one leapt down straight as a rod for three or four hundred feet, preserving its integrity to the bottom; the other formed a curious curve; and both disappeared in a dark chasm, from which issued a rainbow-wreathed cloud of spray. A great curtain of the purest snow hung over the brow of the rock where they both came in sight, and the blue of the sky above it was wonderfully quiet and intense from the contrast. Altogether there was something so spirit-like and ethereal in the source and destiny of these twin waterfalls, issuing apparently from the same snow-wreath far up, and vanishing in the same rainbow-tinted cloud of spray, that we were quite lost in admiration of the sight, and thought this of itself a sufficient recompense of our excursion. On the banks of one of the twin-streams, a long way beyond the precipice, there is a mountain-farm, called Floren, whose loneliness and dreariness must be uncommon even in Norway. Another farther down is called Lia. How the inhabitants get out of the place and into communication with their nearest neighbours is to me incomprehensible. The path must be as dreadful as that of the "Dead Man's Ride" in Vettie-gial. Looking back, when we had advanced about a mile on the lake, the scene was truly extraordinary. The rocks had come together and closed up the entrance, so that we were surrounded on every side by vertical precipices, and there seemed no outlet. It required little exercise of imagination to picture ourselves the ghostly crew of Charon sailing over the Stygian pool. There was something truly infernal in the look of the place, from which a warm human heart recoiled. Dante and Doré might have felt at home in it, but our tamer spirits craved for something less terrific and more earthly. The sun was shut out by the overhanging rocks, and the light was therefore dim and feeble. We were chilled to the marrow by the cold air of the water; and when the clouds gathered, and a heavy shower fell, increasing the sublimity of the scene, the climax of our discomfort was reached. I would advise future visitors to take with them, for this part of the way, a plentiful supply of rugs, for the temperature, even on the hottest day, is like that of the Arctic regions. I know not if there be any superstitious legends connected with this fearful lake. If not, there should be; for I cannot picture a more appropriate haunt for those strange beings, half human and half spiritual, which, according to the Northern mythology, infest the dark fathomless fjords, and require to be appeased every year by the drowning of one or more human victims. It seemed easy, in such a place, to understand how the wildest tales of spirits and monsters of the deep originated. It would be almost impossible to live in Norway and not be superstitious. The powers of nature are so terrible, and on so grand a scale, that they could not fail to be personified and invested with a dread control over human life.

Turning the corner of a great dripping promontory that rose straight from the water into the clouds, like a castle of Thor, a sight burst upon us which for a minute or two nearly took away our breath. It was the Skjeggedal-foss at last! This first glimpse of it was one of those climaxes of life which contrast strangely with its usual tameness and monotony, and make us wonder at the suddenly revealed greatness of our being. There before us was the jealously-guarded secret of the gorge, of which every object all the way had been conscious—the fierce yet beautiful Pythoness of this inmost shrine of nature. As if by one consent the men paused upon their oars, and we gazed in silence. To utter our admiration while that mighty tongue was pouring out its mystic secrets to the trembling rocks we felt would be sacrilege. All waterfalls have a strong family likeness, and should therefore be left undescribed for the imagination to sketch. This one, however, had some peculiar features. The body of water was enormous, and the height upwards of 580 feet. It fell sheer down from the edge of the precipice without touching the rock; and though a great quantity of vapour was disengaged, the vast mass of its waters reached the bottom entire with a solid sound like the fall of a great avalanche. We were upwards of a mile from it, but even at this distance the noise was so penetrating, so transfixing, that the roll of thunder, or the firing of artillery, can give no idea of its fulness and solemnity. As we drew nearer the cataract increased in size and sublimity; while the rocks literally overhung the water. The summits of those on the left were broken up into the most fantastic outlines—rude resemblances of monks, sphinxes, and castles, some of which were half-detached and seemed ready to topple down. Great patches of snow lay wedged in the shady recesses, and increased the peculiarly grey weather-beaten look of the precipices. No more venerable rocks than these bold gigantic masses of gneiss and mica-schist can be found in the world. They are like exposed portions of the skeleton of the earth; and one feels, in looking at them, the appropriateness of the title, "*Ældgamle Rige*," "primeval kingdom," given to their native country by the Norwegian poets.

We landed on an extensive sloping bank lying along the foot of the rocks beside the waterfall. This bank was covered with straggling dwarf birches, and yielded a rich crop of grass wherever there was a clear space of soil among the great lichen-covered boulders. It was evidently a saeter, for there were two or three ruinous wooden sheds erected on it for storing hay until carried down by boat to the farm, and several of those curious wooden frames for drying it were scattered about. In the shallow inlet where we moored our boat, the bottom was composed entirely of thin round pieces of mica-schist, all of the same size, and so like coins that we offered a handful of them playfully to Lars as *sma penge for ein mark*. They had evidently been coined in the mint of the waterfall. I gathered several very rare lichens and mosses among the stones. Nothing could exceed the variety and richness of the flowers growing in the more sheltered places. It was a curious combination of plants which in this country are never seen together. Lowland and Alpine species bloomed side by side without any incongruity. Bluebells, pansies, marsh-marigolds, lilies of the valley, ragged robins, displayed their familiar charms in loving sisterhood with the shiest beauties which in Britain are found only in one or two isolated spots among the loftiest Highland mountains. *Ajuga alpina, Bartsia alpina, Salix reticulata* and *herbacea, Pedicularis lapponica, Cornus suecica, Rubus arcticus, Smilacina bifolia, Saxifraga cernua* and *rivularis, Thalictrum alpinum, Pinguicula villosa, Sonchus alpinus, Cerastium alpinum, Ranunculus glacialis, Hierochloe borealis, Phleum alpinum;* these and many more Alpine plants, exceedingly rare in Britain, were gathered on this little plot of ground. Here, as on the summits of the Highland mountains, the *Silene acaulis* formed great soft carpets on the mossy ground, with its tufted foliage hardly seen for the multitude of rosy blossoms. The wondrous loveliness of the large blue eyes of the Alpine *Veronica* —looking out upon me from behind the shelter of every stone—haunts me still. And high on the tops of the largest boulders the magnificent *Saxifraga cotyledon* waved its long rich spike of snowy blossoms in every gust of wind. It is well named *Berg-kongen*, "king of the rocks," for it is a truly royal plant. It recalled many a delightful memory of the Alps, where I gathered it among the grandest scenes. I could have spent a whole day botanizing in this rich habitat; but as our time was limited, I was obliged to content myself with the species that came most readily to hand, leaving many a rare and beautiful plant "to blush unseen, and waste its sweetness on the desert air."

Wrapped in our waterproofs, we climbed among the wet rocks, past the limits of vegetation, as near as we could venture to the edge of the abyss; and there through a dense writhing mist of spray, which poured in streams from our garments, we caught a glimpse of a huge wreath of snow lining the sides of the caldron all round, which seems to be perpetual. Into the heart of this cloven wreath the cataract fell with an appalling sound, and from thence plunged down in a series of smaller falls into the lake. We could not see the nature of the linn beneath the cataract, for it was filled with blinding vapour, which rushed half-way up the sides of the black rocks and fell down again in numberless cascades—which of themselves would have attracted admiration in any other place. High overhead on the sky-line the vast volume of water burst over the ledge of rock. We watched it descending, churned and ground by the concussion into the smallest atoms, and yet forming in their aggregate mass a snowy pillar of gigantic dimensions and irresistible strength. We lingered on the spot, loth to leave, fascinated by the indescribable wildness and terror of the sight; and when we did go, we looked behind again and again, for the eye was not satisfied with seeing. We rowed safely back to the farm, where we had the rare luxury of paying a landed proprietor a sum equivalent to two shillings and sixpence of English money, and receiving in acknowledgment of our munificence a hearty shake of the hand and "*mange tak*" (many thanks). The steep staircase was ascended with less trepidation than it was descended: and over the bare house-roof of rock we walked with greater boldness, in the erect attitude that becomes a man; having, at the guide's suggestion, taken the precaution of putting off our shoes, and going across in our stockings. All the way as we descended we obtained through the trees magnificent views of the snowy plateau of the Folgefond, reddened on its highest part by the exquisite *abendglühen*, or after-glow of sunset. We reached Odde at eight o'clock, moderately fatigued and immensely gratified with our excursion, but leaving the comparative merits of the Voring-foss and the Skjeggedal-foss an open question, to be settled for himself by each tourist who follows in our footsteps.

CHAPTER VI.

THE PASS AND HOSPICE OF THE GREAT ST. BERNARD.

THERE is no episode in continental travel more interesting at the time, and more suggestive of pleasing memories afterwards, than a visit to the Great St. Bernard Hospice. It does one moral as well as physical good. The imagination is stimulated by the associations of the place, and the heart filled with the feverish unrest and love of excitement so characteristic of the present age is rebuked and calmed by the loneliness and monotony of the life. Every one has heard of its dogs and monks, and its travellers rescued from the snow-storms. Pictures of it used to excite our wonder in the days of childhood; descriptions of it in almost every Swiss tourist's book have interested us in maturer years; while not a few of us have made a pilgrimage to the spot, and thus given to the romantic dreams and fancies of early life a local habitation and a name. Still, trite and worn-out as the subject may appear, it is impossible by any amount of familiarity to divest it of its undying charm; and those who have visited the scene, so far from their interest in it being exhausted, have only been made more enthusiastic in its favour, and more anxious to compare or contrast their own experience with that of every new traveller who writes upon it. To the botanist especially the region is exceedingly interesting. In ascending the pass he has an opportunity of noticing the various types of vegetation that occur in the different zones of altitude, from the plants of Southern Europe in the valleys to the Arctic flora below the line of perpetual snow. There are few places where so great a variety of Scandinavian forms may be gathered as on this crest of the Pennine Alps, growing among forms that are peculiar to the locality. Even the unscientific traveller is struck with their extreme luxuriance and beauty. They form an essential feature in the landscape, which the most careless will notice and remember with pleasure long afterwards, associating the beds of lovely Alpine plants with the fresh, bracing air, the bright rejoicing waters, and the noble prospects of the mountain heights.

About the beginning of August, two years ago, I had the pleasure of visiting this celebrated spot in company with two friends. We set out early in the morning in a char-à-banc, or native droskey, drawn by a mule from the "Hôtel Grande-Maison-Porte," at Martigny, the Roman Octodurus, and the seat of the ancient bishops of Valais. This is a low, damp, uninteresting place, much infested with

S

a small, black gnat, whose sting is very painful, bred in the marshes of the Rhone. Being a capital centre of excursions to Lago Maggiore over the Simplon, to Aosta and Turin over the St. Bernard pass, and to Chamouni by the Tête-Noire, or the Col de Balme, it is exceedingly gay and animated every evening during the summer, owing to the arrival of tourists, and desolate and deserted every morning, owing to their departure. The sun was shining with almost tropical heat, rapidly ripening the walnuts along the avenues of the town, and the grapes hanging in rich profusion on the trellises of the houses; the sky was without a cloud, and everything promised a delightful trip. Passing through a small unsavoury village called Martigny le Bourg, our route crossed the Dranse by a substantial bridge; and at a little distance beyond a guide-post indicated to the right the way to Chamouni, and to the left to St. Bernard. The entrance by the pass of the Dranse is magnificent, reminding us, though on a grander scale, of the mouth of Glenlyon in Perthshire. Lofty slopes, and precipices richly wooded, approached from both sides so closely that there was hardly room left for the passage of the powerful stream, which, turbid with glacier mud, roared and foamed over enormous blocks of stone. The road, without parapet or railing, overhung the river, and in one place was carried through a tunnel called the *Gallerie Monaye*, upwards of two hundred feet long, cut out of the solid rock. We passed through scattered villages sweetly embosomed among walnut and chesnut trees, but presenting many saddening signs of the poverty and wretchedness of the inhabitants. An unusually large proportion of the people were afflicted with goitres, and here and there we saw sitting on the thresholds of their dirty chalets loathsome cretins, basking in the sun, whose short, shambling figures and unnaturally large round heads and leering faces afflicted us amid the beauty of nature around like a nightmare. The ground was everywhere most carefully cultivated. Every particle of soil among the rocks, however scanty or steep, was terraced up with walls, and made to yield grass, corn, or potatoes. High up on the brink of precipices that seemed almost inaccessible, bright green spots indicated the laborious care of the peasantry; and to these, as soon as the winter snows disappeared, sheep were carried up every year, one by one on men's backs, and left there till the end of summer, when they were carried down, considerably fattened, in the same picturesque fashion. The lower meadows by the roadside were exceedingly beautiful, of the most vivid green, covered with myriads of purple crocuses and scarlet vetches, and murmurous with the hum of innumerable grasshoppers. Gay butterflies, and insects of golden and crimson hues, never seen in this country, flitted past in the warm sunshine; and the fragrance of the Arolla pines filled all the air with a highly stimulating aromatic balm. As it was the festal day of the "Assumption of the

Virgin," one of the grandest fêtes of the Roman Catholic Church, groups of peasants,—the men dressed in the brown cotton blouses peculiar to the district, and the women wearing a curious head-dress consisting of a broad tinselled ribbon plaited and set on edge round a cap, each carrying her prayer-book in her hand, wrapped in a white pocket-handkerchief,—passed us on their way to the chapel at Martigny. On all sides we noticed exceedingly distinct traces of two great natural phenomena which had overwhelmed the district, separated from each other by thousands of years. Almost every exposed rock was polished and striated by ancient glaciers; and the granite boulders, which they had brought down with them, were seen perched upon the schist and limestone precipices hundreds of feet above the river. The whole valley from St. Bernard to Martigny, with its tributary glens, must have been the channel of a vast system of glaciers descending from the crest of the Pennine Alps during the glacial epoch, when all the glaciers of Europe and Asia were far more extensive than they are now. The other phenomenon to which allusion has been made was also caused by glacier action, but of a different kind. In one of the narrow side gorges of the valley, called the Val de Bagne, there is a glacier known as the Glacier de Getroz, which hangs suspended over a cliff five hundred feet high. The end of this glacier is continually breaking off, and falling over the precipice into the bottom of the gorge, where the fragments of ice accumulate and form enormous heaps. In the year 1818 these fallen masses had been piled up to an unparalleled extent, and choked up the narrow, vault-like outlet of the gorge. Behind this icy dam the water of the east branch of the Dranse increased, until at length a lake was formed, nearly a mile long, a quarter of a mile wide, and about two hundred feet deep. The inhabitants of the valley watched anxiously the gradual rise of the waters, knowing that when the warm season should come the icy bank would melt, and the reservoir be at once discharged. Many of them fled in the spring, with their goods and cattle, to the higher pasturages. A tunnel, seven hundred feet long, was cut into the ice, which gradually let off a considerable part of the water without any damage. But a hot June sun and the warmth of the water so gnawed into the ice that on the afternoon of the 16th of the month the barrier burst all at once, and a prodigious mass of water, upwards of five hundred and thirty millions of cubic feet, rushed down the valley with fearful fury, carrying everything before it, and marking its course all the way to the lake of Geneva, fifty miles distant, with gigantic ruins. Many lives were lost, and property to nearly the value of a million sterling was destroyed. To prevent a repetition of this awful calamity, for a similar event occurred in 1595, and the same cause is still in operation, spring water is led by means of a long wooden trough to the dam of ice formed by the falling fragments of the glacier; and the warmth of this water cuts like a saw the ice as soon as deposited, and thus cleaves a passage for the river and prevents its waters from accumulating. The autograph of this tremendous inundation was written, like the mystic "Mene, mene," of Belshazzar's palace, in the huge stones in the bed of the river, and in the gravelly and stony spots far up the sides of the valley, mingling with the relics of ancient glacier action, but easily distinguishable from them.

Passing through Sembranchier, a picturesque village, with the ruins of an enormous castle of the Emperor Sigismund on a hill in its vicinity, and Orsières, situated at the junction of the valleys of Ferret and Entremont, distinguished by a very ancient tower rising high above its curious houses, the road ascended by a series of well-executed zigzags through a rich and highly-cultivated country to Liddes. Deep down among wild rocks the Dranse pursued its turbulent course unseen, revealing its presence only by an all-pervading murmur in the air. The view extended over an undulating upland landscape of green fields, diversified by wooden frames for drying the corn, somewhat like the curious structures for drying hay to be seen on Norwegian mountain farms. The huge summit of Mont Velan, 12,000 feet high, formed the most conspicuous object on the horizon before us, its dark rocks contrasting finely with its dazzling snows and the rich fields of deep blue sky above it. A cool breeze blew down upon us from the snowy heights, and was inexpressibly refreshing after the stifling heat of the valley. About four o'clock in the afternoon we came to a strange old village, called St. Pierre,—the last on the route,—situated on a kind of plateau, about 5,000 feet above the level of the sea. It was a very dirty, miserable place; and we were victimized by the innkeeper of the Hôtel au Dejeuner de Napoléon, having been charged fifteen francs for a blue scraggy chicken, not much larger than a sparrow, a plate of potatoes fried in rancid grease, and a bottle of Beaujolais wine as sour as vinegar. A remarkably quaint old church, built in the tenth century, still exists in the village. A tablet with a Latin inscription by Bishop Hugo of Geneva, the founder of the church, commemorates a victory obtained by the inhabitants over the Saracens, who had ravaged the district with fire and sword. A Roman milestone is also built into the wall of the enclosure near the tower. In modern times the place is chiefly interesting as being one of the resting-places of Napoleon in his passage over the Alps, and the birthplace of his famous guide. A little beyond it there is a deep gorge with a splendid, full-bodied waterfall, which we visited. The sides of the pools and the sloping banks were fringed with clusters of tall monkshood, whose blue flowers mingled with the snowy foam of the water; while the large yellow flowers of the Swiss foxglove (*Digitalis grandiflora*) peeped out with a very brilliant effect among the bushes. Across the gorge, a frail bridge, with an arched gateway, constructed by Charlemagne, gave access to the main road, which led through the forest of St. Pierre in the Defilé de Charreire, and was cut in many places out of the solid rock. Below us, at the foot of perpendicular precipices several hundred feet in depth, the Dranse, still a powerful stream, formed innumerable foaming cascades. There was no wall or abutment to protect us. The off-hand wheel of the conveyance was always within a foot of the edge. I was sitting on the side nearest the precipice, and often could have easily let fall a stone from my hand right down into the river. The least false movement on the part of the driver would inevitably have hurled us over to destruction. And yet we went safely and pleasantly along at full speed, our hearts now and then, when we came to a more trying place than usual, perhaps a little higher than their normal position. It was in this defile of Charreire that Napoleon encountered his most formidable difficulties. The old road was declared by Marescat, chief of the engineers, as "barely passable for artillery." "It is possible! let us start then!" was the heroic reply of his master. It was a favourite maxim with him that wherever two men could set foot an army had the means of passing; and he acted upon this maxim on this occasion. As it was about the end of May, the snows were melting fast, and thus greatly increased the dangers and difficulties of the route. "The artillery carriages were taken to pieces and packed on mules; the ammunition was also thus transported; whilst the guns themselves, placed on the trunks of trees hollowed out, were dragged up by main strength,—a hundred soldiers being attached to each cannon, for which laborious undertaking they received the sum of 1,200 francs. At the Hospice each soldier partook of the hospitality of the brethren."

In about an hour and a half we came to a solitary inn, called the Cantine de Proz, kept by a man of the name of Dorset, who is very civil to travellers. No other dwelling was in sight. A number of diminutive cows wandered about on the short smooth turf, bright with the lovely Alpine clover; the sweet tinkling of their bells, combined with the monotonous sighing of the infant Dranse, giving us a lonely and far-away feeling, as if we had reached the end of the world. A corner of the Glacier de Menouve, of dazzling whiteness, appeared in sight, far up among stern precipitous rocks, of a peculiarly bald and weather-worn appearance. Above the cantine, a little plain, called the Plan de Proz, about 5,500 feet above the sea, sloped up, seamed in every direction with grey watercourses, but gemmed with innumerable brilliant clusters of the snowy gentian. Leaving our conveyance at the inn, and taking with us the mule and the driver as a guide, we set off on foot across the plain, to the entrance of a kind of gorge, called the Defilé de Marengo, which is exceedingly steep and difficult of ascent. A considerable stream, confined within narrow bounds, roars and foams within a few feet of the pathway, so that in wet weather its swollen waters must render the defile impassable. Among the rocks, wherever any particles of soil lodged, rich cushions of moss spread themselves, wild auriculas nestled in the crevices, and large patches of crowberry and blaeberry bushes crept over the boulders. These blaeberry bushes fringed the pathway up to within a short distance of the Hospice; and nowhere in Scotland have we seen the fruit so plentiful or so large and luxurious. Basketfuls could be gathered in a few minutes without diverging more than a yard or two from our course; and yet it seems never to be touched. The sides of the stream were decked with the large woolly leaves and brown flowers of the Alpine *Tussilago*, which takes the place at this elevation of the common butter-bur, whose enormous umbrella-like leaves form such a picturesque adornment of lowland rivulets. After an hour's stiff ascent, we came to two ruinous-looking chalets, built of loose stones, one of which served as a place of refuge for cattle, while the other was the old morgue, now used as a shelter-place for travellers, where they wait, if overtaken by storms, till the servants of the monastery come down with a dog to their rescue, which they do every morning when the weather is unusually severe. They bring with them on such occasions wine and provisions to restore the exhausted and half-frozen traveller; and guided by the faithful dogs, who alone know the way,—thirty feet of snow being not unfrequently accumulated in the worst parts of the pass,—they are all brought safely to the hospitable shelter of the convent. From this point the defile receives the ominous name of the Valley of Death; and the track is marked by tall, black poles, and here and there by a cross, marking the scene of some tragic event. Within a short distance of the Hospice, an iron cross commemorates the death of one of the monks who perished on that spot by an avalanche in November 1845. Between these grim memorials of those to whom the place had been indeed the valley of the shadow of death we toiled up the rough and arduous path, panting and perspiring, greatly aided by our alpenstocks. For my own part, I thought the way would never end. I turned corner after corner of the defile, but still no trace of human habitation. My knees were about to give way with fatigue, the rarity of the air was making itself known to me in thirst and headache, my pulse had advanced from 60 beats at Martigny to 83 at this elevation, and I would gladly have rested awhile. But the shades of night were falling fast, so the banner with the strange device had still to be unfurled. I had in my own experience during this ascent a more vivid conception than I could otherwise have realized of the feverish longing which the lost wanderer in the snow has for a place of refuge and rest. If I, a mere summer tourist, bent upon reaching the Hospice only to gratify a love of adventure, and to realize a romantic sensation, had such a desire, how much more ardent must be the longing of the poor traveller, overtaken by the dreadful *tourmente*, blinded and benumbed by the furious drift, to whom reaching the Hospice is a matter of life and death! At last, at the very summit of the pass, I saw the Hospice looming above me, its windows glittering in the setting sun. Fatigue and weariness all forgotten, I eagerly clambered up the remaining part of the ascent, along a paved road overhanging a precipice, and in a few minutes stood beside the open door. At first I could hardly realize the fact that the convent, about which I had read so much, which I had so often seen in pictures and pictured in dreams, was actually before me. It had a very familiar look, appearing exactly as I had imagined. I did not approach it in the orthodox fashion,—exhausted and half-frozen amid the blinding drifts of a snow-storm, and dragged in on a dog's back! On the contrary, the evening was calm and summer-like; the surrounding peaks retained the last crimson blush of the exquisitely beautiful *abend-gluhen*, or after-glow of sunset; the little lake beside the convent mirrored the building on its tranquil bosom; the snow had retreated from the low grounds, and only lingered on the lesser heights in the form of hardened patches wedged in the shady recesses of the rocks. I could not have seen the place under more favourable auspices; and yet, nevertheless, the scene was inexpressibly forlorn and melancholy. There was an air of utter solitude and dreariness about it which I have never seen equalled, and which oppressed me with a nameless sadness. There was no colour in the landscape,—no cheerful green, or warm brown, or shining gold, such as relieves even the most sterile moorland scenery in this country. Everything was grey and cold — the building was grey, the rocks were grey, the lake was grey, the vegetation was grey, the sky was grey; and when the evening glow vanished, the lofty peaks around assumed a livid ghastly hue, which even the sparkling of their snowy drapery in the first beams of the moon could not enliven. Not a tree, not a shrub, not even a heather bush, was in sight. It seemed as if Nature, in this remote and elevated region, were dead, and that I was gazing upon its shrouded corpse in a chamber draperied with the garments of woe.

The Monastery itself is a remarkably plain building, destitute of all architectural pretensions. It is in fact a huge barn, built entirely for use and not for elegance. It consists of two parts—one fitted up as a chapel, and the other containing the cells of the monks, and rooms for the accommodation of travellers, divided from each other by whitewashed wooden partitions. It is built in the strongest manner,—the walls being very thick, and the windows numerous, small, and doubly-glazed, so as most effectually to withstand the fearful storms of winter. There is a small separate building on the other side of the path, called the Hôtel de St. Louis, which is used as a granary, and as a sleeping-place for beggars and tramps. It also provides a refuge in the case of fire, from which the Hospice has frequently suffered severely, being on two occasions nearly burnt to the ground. Ladies were formerly entertained in this building, as it was deemed out of place to bring them into the Monastery. But these scruples have now been overcome, and ladies are freely admitted to all parts of the place, and allowed to sleep in the ordinary rooms. The monks of the present day have not the same dread of the fair sex which their patron saint is said to have cherished. Indeed, the good fathers are particularly delicate and profuse in their attentions to ladies, giving to them the best places at table, and serving them with the choicest viands. In fact, the company of ladies is one of the best letters of introduction that a party can bring with them; for though the monks are proverbially kind and attentive to all persons without distinction, and especially considerate, from a sympathetic feeling, towards bachelors, yet if they have a warmer place than another in their hearts, it is reserved for lady travellers; and who would blame them for it?

The St. Bernard Hospice is the highest permanent habitation in Europe, being 8,200 feet above the level of the sea, or nearly twice the height of Ben Nevis. There are, indeed, several chalets in the Alps that are still higher, but they are tenanted only during the three summer months, when the people employ themselves in tending goats and manufacturing cheeses from their milk. About the end of September they are deserted, and the shepherds descend to the valleys. The severity of the climate at the Hospice is so great, that the snow never leaves the level ground for nine months in the year. Snow showers are almost always falling, even in the mildest weather; and there are scarcely three successive days in the whole twelve months free from blinding mists and biting sleet. The mean temperature is 30° Fahr., exactly that of the South Cape of Spitzbergen. In summer it never exceeds 48°, even on the hottest day; and in winter, particularly in February, the thermometer not unfrequently falls 40° below zero,—a degree of cold of which we in this country can form no conception. What greatly increases the severity of the climate is the fact that the Hospice is situated in a gorge pierced nearly from north-east to south-west, in the general direction of the Alps, and consequently in the course of the prevailing winds; so that, even in the height of July, the least breath of the *bise*, or north wind, sweeping over the lofty snow region, always brings with it a degree of cold extremely uncomfortable. The effect of this bitter Arctic climate upon the monks, as might be expected, is extremely disastrous. The strongest constitution soon gives way under it. Headaches, pains in the chest and liver, are sadly common. Even the dogs themselves, hardy though they are, soon become rheumatic and die. Seven years is the longest span of their life, and the breed is with the utmost difficulty kept up. All the monks are young men, none of them having the grey hair, and long venerable beard, and feeble stooping gait, which are usually associated with the monastic fraternity. In fact, the intensity of the climate prevents any one from reaching old age. The prior, M. Joseph de l'Eglise, has been longer in the convent than any other monk, having spent there considerably more than the half of his life. But though only forty-six years of age, he looked a withered, pinched old man, suffering constantly and acutely from the disorders of the place, yet bearing his illnesses in patient uncomplaining silence, and going about his work as though nothing were the matter with him. The monks begin their noviciate, which usually lasts about fourteen years, at the age of eighteen; but few of them live to complete it. The first year of residence is the least trying, as the stock of health and energy they have brought with them enables them successfully to resist the devitalizing influence of the monotonous life and the severe climate; but every succeeding year they become less and less able to bear the cold and privations, and they go about the convent the ghosts of their former selves, blue and thin and shivering. Before they have succumbed, they go down to the sick establishments in the milder climate of Martigny or Aosta, or they serve as parish priests in the different valleys around. But, in many cases, this remedy comes too late. They perish at their posts, literally starved to death. The annals of the convent contain many sad records of such devotion;

and they thrill the heart with sympathy and admiration.

We mounted the stair in front of the door of the Hospice, and entered, preceded by our guide. In the wall of the vestibule we noticed a large black marble tablet, bearing the following inscription in gilt letters: "Napoleoni I. Francorum Imperatori, semper augusto Reipublicæ Valesianæ restauratori, semper optimo Ægyptiaco, bis Italico, semper invicto, in monte Jovis et Sempronii semper memorando respublica Valesiæ grata, 2 Dec. 1804." At the top of a short flight of steps, our guide rang a large bell twice, and immediately a door opened and a polite and gentlemanly monk appeared, dressed in a long black coat with white facings, and with a high dark cap, similarly decorated, upon his head. He welcomed us with much politeness, and beckoning us to follow him, conducted us through a long vaulted corridor, dimly lighted by a solitary lamp, where the clang of an iron gate shutting behind us, and the sound of our own footsteps on the stone floor, produced a hollow reverberation. He brought us into a narrow room, with one deeply-recessed window at the end, containing three beds simply draped with dark crimson curtains, and all the materials for a comfortable toilet. There are about eighty beds for travellers of better condition in the monastery, and accommodation for between two and three hundred persons of all classes at one time. Speedily removing our travel-stains, we rejoined our host in the corridor, who showed us into the general reception room, where we found lights and a smouldering wood fire upon the hearth. The walls of the room, lined with pine wainscot, were hung with engravings and paintings, the gifts of grateful travellers; while in one corner was a piano, presented by the Prince of Wales shortly after his visit to the Hospice. Two long tables occupied the sides, covered with French newspapers and periodicals, among which we noticed several recent numbers of *Galignani* and the *Illustrated London News*. We went instinctively at once to the fire, but found it monopolized by a party of Italians and Germans, who showed no disposition to admit us within the magic circle. We elbowed our way in, however, and had the satisfaction of crouching over the smouldering logs with the rest, and admiring the beautifully-carved marble mantelpiece. One of the monks very considerately came in with an armful of wood and a pair of bellows, and, replenishing the fire, speedily produced a cheerful blaze, which thawed us all into good-humour and genial chattiness. We felt the cold exceedingly; the thermometer in one of the windows of the room registering six degrees below the freezing point. At Martigny, in the morning, the thermometer indicated about eighty degrees in the shade; so that in less than twelve hours we had passed from a tropical heat sufficient to blister the skin exposed to it to an Arctic cold capable of benumbing it with frost-bites. The rooms of the convent are heated all the year round; and at what an expense and trouble it may be judged, when the fact is mentioned, that every particle of the fuel consumed is brought on the backs of mules over the Col de Fenetre, a continuous ascent of nine thousand feet, from the convent forest in the valley of Ferret, twelve miles distant. Water, too, boils at this elevation at about 187° Fahr., or twenty-five degrees sooner than the normal point; and in consequence of this it takes *five* hours to cook a piece of meat, which would have taken only three hours to get ready down in the valleys, and a most inordinate quantity of fuel is consumed in the kitchen during the process. The most essential element of life in this terrible climate is yet, sad to say, too rare and precious to be used in sufficient quantity. What would not the poor monks give for a roaring blazing coal fire, such as cheers in almost limitless measure our homes on the winter nights, when they sit shivering over the dim glimmer of a wood fire carefully doled out in ounces!

Having arrived too late for supper, which is usually served at six, the dinner hour being at noon, an impromptu meal was provided for us and the other travellers who were in the same position. Though hastily got up, the cooking of it would have done credit to the best hotel in Martigny. It consisted of excellent soup, roast chamois, and boiled rice and milk, with prunes. A bottle of very superior red wine, which was said to be a present from the King of Sardinia, was put beside each person; and a small dessert of nuts and dried fruits wound up the entertainment. The Clavandier presided, and by his courteous manners made every one feel perfectly at home. The conversation was carried on exclusively in French, which is the only language spoken by the fathers. Coming in contact during the summer months with travellers from all parts of the world, and devoting the long winter to hard study, in which they are helped by the superior, who is a man of great learning, the monks are exceedingly intelligent, and well acquainted with the leading events of the day, in which they take a deep interest. Some of them are proficients in music; others display a taste for natural history; and they all combine various accomplishments with their special study of theology and the patristic literature. They are also very liberal in their views, having none of the bigotry and intolerance which we usually associate with the monastic order, and which is so conspicuous in the curés of the Papal Swiss cantons. A striking example of this was related to us at the time. A week before our arrival, an Episcopalian clergyman, happening to be staying with a party of Englishmen in the Hospice on a Sunday, asked permission of the superior to conduct a religious service with his countrymen in the refectory. This was not only granted with the utmost cordiality, but the chapel itself was offered to him for the purpose, which offer, however, he declined in the same spirit in which it was made, unwilling to trespass to that extent upon their catholicity.

Conversing pleasantly on various subjects with our host and the guests around, we did ample justice to the good cheer. Fridays and Saturdays, we understood, were fast days; but though the brethren fasted, no restriction was put during those days upon the diet of travellers—the table being always simply but amply furnished. The task of purveying for the Hospice, which falls to the Clavandier, is by no means an easy one, when it is considered that upwards of sixteen thousand travellers, with appetites greatly sharpened by the keen air, are entertained every year; and not a single scrap of anything that can be eaten grows on the St. Bernard itself. All the provisions, which must consist of articles that will keep, are brought from Aosta, and stored in the magazines of the convent. During the months of June, July, and August, when the paths are open, about twenty horses and mules are employed every day in carrying in food and fuel for the long winter. The country people also bring up gifts of cheese, butter, and potatoes, in gratitude for the kind services of the monks. Several cows are kept in the convent pastures on the Italian side, and their milk affords a grateful addition to the food of the monks. During winter they have no fresh meat at all, being obliged to subsist upon salt beef and mutton, usually killed and preserved in September; and what is still worse, they have no vegetables, all attempts at gardening in the place having proved abortive; so that not unfrequently scurvy is added to their sufferings.

After an hour or two's chat around the fire, and a very cursory but most interesting inspection of the pile of visitors' books, which contain many celebrated names, and a great deal that is curious and admirable in the way of comment upon the place, our host bade us all good-night, and I too was very glad to retire. A bright moon shone in through the curtainless window of my bedroom, and lay in bars on the bare floor. Outside the view was most romantic, the moonshine investing everything, snowy peaks, jagged rocks, and the bare terraces around, with lights and shadows of the strangest kind. A pale blue sky, spiritual almost in its purity and transparency, in which the stars glimmered with a cold clear splendour, bent over the wild spot; and the loneliness and silence were unlike in their depth and utterness anything I had ever before experienced. Snatching, like Gray's schoolboy, a few minutes of fearful joy from the contemplation of the weird scene, worn-out nature summoned me to bed. There was a perfect pile of blankets and a heavy down quilt above me, under which I lay squeezed like a cheese in a cheese-press, but I utterly failed to get warm. Sleep would not be wooed. I lay and watched the shadows on the floor, and thought of many unutterable things, and wondered at the strange vicissitudes of life, which so often place us unexpectedly in situations that were the ideals of our youth. About five o'clock in the morning, just as the grey dawn was stealing in, I was thoroughly roused from a dozing, semi-torpid state, into which I had sunk, by the ringing of the convent bell for matins; and shortly afterwards the rich tones of an organ, mellowed by the distance, pealed from the chapel with an indescribably romantic effect. I arose and dressed with chattering teeth, and then went out into the raw air. I walked beside the small, desolate-looking lake beside the Hospice, where never fish leaped up, and on which no boat has ever sailed. Being the highest sheet of water in Europe, fed by the melting of the snows, it is frequently frozen all the summer; and when thawed, it lies " like a spot of ink amid the snow." Passing a pillar at the end of the lake, and a curious heraldic stone beside a spring, I had crossed the boundary between Switzerland and Piedmont, and was now in Italy. Climbing up the bare rocks to a kind of esplanade, near a tall cross inserted in a massive pedestal of chlorite-schist, and bearing the inscription, "Deo Optimo Maximo," which guides the traveller from the Italian side of the pass to the convent, I sat down and surveyed the scene. The snowy dome of Mont Velan filled up the western horizon. On my left the gorge was shut in by the rugged range of Mont Mort, Mont Chenaletta, and the Pic de Dronaz. Below me I could see, through the

writhing mist, glimpses of the green corrie, called
"La Vacherie," where the cattle of the Hospice
grazed under the care of a few peasants, whose
wretched chalets were the only habitations: while
beyond, to the southward, rose up a strange Sinai-
like group of reddish serrated rocks, entirely
destitute of vegetation, with wreaths of dark cloud
floating across their faces, or clinging to their
ledges, and greatly increasing their savage gloom.
An air of utter desolation and loneliness pervaded
the whole scene. No sounds broke the stillness,
save such as were wonderfully congenial with the
spirit of the place, the sighing of the wind as it
ruffled the surface of the lake, the occasional tinkle
of the cow-bells far below, the deep baying of the
St. Bernard dogs, or the murmur of a torrent far
off, that came faint and continuous as music heard
in ocean shells.

I had ample evidence around — if my dripping nose and icy hands did not convince me—
of the extreme severity of the climate. The
vegetation was exclusively hyperborean, exactly
similar in type to that which flourishes around
the grim shores of Baffin's Bay. I had gathered
the same species on the summits of the highest
Scottish mountains, and afterwards on the Dovrefjeld in Norway. The reindeer moss of Lapland whitened the ground here and there, interspersed with a sulphur-coloured lichen which
grows sparingly on the tops of the Cairngorm
range. Large patches of black *Tripe de Roche*—

the lichen which Sir John Franklin and his party, in the Polar regions, were once, in the absence of all other food, compelled to eat, along with the remains of their old shoes and leather belts—clung to the stones, looking like fragments of charred parchment; while an immense quantity of other well-known Arctic lichens and mosses covered the level surface of each exposed rock, as with a crisp shaggy mantle, that crunched under the foot. There were no tufts of grass, no green thing whatever. Tiny grey saxifrages, covered with white flowers, grew in thick clumps, as if crowding together for warmth, along with brilliant little patches of gentian, whose depth and tenderness of blue were indescribable, and tufts of Aretias and Silenes, starred with a profusion of the most exquisite rosy flowers, as though the crimson glow of sunset had settled permanently upon them. The Alpine forget-me-not, only found in this country on the summits of the Breadalbane mountains, cheered me with its bright blue eyes everywhere; while the "Alpine lady's mantle" spread its grey satiny leaves, along with the Arctic willow, the favourite food of the chamois, over the stony knolls, as if in pity for their nakedness. I found a few specimens of the beautiful lilac *Soldanella alpina*, and also several tufts of the glacier ranunculus, on a kind of moraine at the foot of a hardened snow-wreath. The ranunculus was higher up, and grew on the loose *débris*, without a particle of verdure around it. It seemed like the last effort of expiring nature to fringe the limit of eternal snow with life. Its foliage and flowers had a peculiarly wan and woebegone look. Its appeal was so sorrowful, as it looked up at me, with its bleached colourless petals, faintly tinged with a hectic flush, that I could not help sympathising with it, as though it were a sensitive creature. But the flower that touched me most was our own beloved "Scottish blue-bell." I was surprised and delighted beyond measure to see it hanging its rich peal of bells in myriads from the crevices of the rocks around, swaying with every breeze. It tolled in fairy tones the music of "Home, sweet home." It was like meeting a friend in a far country. It was the old familiar blue-bell, but it was changed in some respects. Its blossom was far larger, and of a deep purple tinge, instead of the clear pale blue colour which it has in this country. It afforded a striking example of the changes which the same plant undergoes when placed in different circumstances. I could see in its altered features modifications to suit a higher altitude and a severer climate. In the Alps all the plants have blossoms remarkably large in proportion to their foliage, and their colours are unusually intensified, in order that they may get all the advantage of the brief but ardent sunshine, so as to ripen their seed as rapidly as possible. And this unprincipled little blue-bell in the vicinity of the Monastery had exchanged the clear blue of the Scottish Covenanter for the purple and fine linen of the Romish hierarchy, and was just like many others, animals as well as plants, doing in Rome as they do in Rome! In this desolate, nature-forsaken spot, where an eternal winter reigns, the presence of these beautiful Alpine flowers, doing their best to make the place cheery, brought a peculiar indescribable feeling of spring to my heart, reminded me irresistibly of the season which is so sad amid all its beauty and promise —the first trembling out of the dark—the first thrill of life that comes to the waiting earth—and then the first timid peering forth of green on hedge and bank; and, like Coleridge's "Ancient Mariner," I said:

> " Oh, happy living things! no tongue
> Their beauty might declare;
> A spring of love gushed from my heart,
> And I blessed them unaware."

It is impossible to gaze on the St. Bernard pass without feelings of the deepest interest. It stands as a link in the chain that connects ancient and modern history—departed dynasties and systems of religion with modern governments and fresh creeds; and in this part the continuity has never been broken. Bare and bleak as is the spot, it is a palimpsest crowded with relics of different epochs and civilizations, the one covering but not obliterating the other. Every step you take you set your foot upon "some reverend history." Thought, like the electric spark, rapidly traverses the thousand historical links of the chain of memory.

You feel as if in the crowded valley in the vision of Mirza. All the nations of the earth—Druids, Celts, Romans, Saracens, French, Italians—seem to pass in solemn file, a dim and ghostly band, before your fancy's eye. Names that have left an imperishable wake behind them—Cæsar, Charlemagne, Canute, Francis I., Napoleon Buonaparte—have traversed that pass. Europe, Africa, and Asia have poured their wild hordes through that narrow defile. The spot on which the convent is erected was held sacred and oracular from time immemorial. Like the Tarpeian rock and the site of ancient Rome, like the stern solitudes of Sinai and Horeb, it had a *religio loci* and a consecrated shrine from the remotest antiquity. The weird, wild aspect of the place gave it an air of terror, and naturally associated it with the presence of some mysterious supernatural being. On a little piece of level ground near the lake, called the Place de Jupiter, on which the ruinous foundations of an ancient Roman temple may still be seen, a rude altar, built of rough blocks of stone, was erected three thousand years ago, and sacrifices offered on it to *Pen*, the god of the mountains, from whom the whole great central chain of Switzerland received the name of Pennine Alps. The custom of building cairns on the highest points of our own hills is supposed to have been derived from the worship of this divinity, which seems at one time to have spread over the whole of Europe. The names of many of the Highland mountains bear significant traces of it. Ben Nevis means "Hill of heaven," and Ben Ledi signifies "Hill of God," having near the summit some large upright stones, which in all probability formed a shrine of the god Pen, whose Gaelic equivalent, as Beinn or Ben, has been bestowed on every conspicuous summit. Who the primitive people were that first erected the rude altars on the St. Bernard pass to their tutelary deity, we know not. They may have been allied to those strange Lacustrines who studded the lakes of Switzerland and Italy with their groups of dwellings, at the time that Abraham was journeying to Canaan, and whose relics, recently discovered, are exciting so much interest among archæologists. They were no doubt Celtic tribes; but, as Niebuhr says, "the narrow limits of history embrace only the period of their decline as a nation." The few fragments that are left of their language, like the waves of the ancient ocean, have a mysterious murmur of their own, which we can never clearly understand.

For hundreds of years this unknown people worshipped their god, and held possession of their territories undisturbed; but the day came when they were compelled to yield to a foreign invader, who fabricated his weapons of iron, and wielded them with a stronger arm. Rome had established a universal supremacy, and sent its conquering legions over the whole of Europe. The stupendous barrier of the Alps offered no obstruction. Through its passes and valleys, led on by Cæsar Augustus in person, they poured like an irresistible torrent, washing away all traces of the former peoples. They demolished the old Druidic altar on the summit of the St. Bernard, and erected on its site a temple dedicated to Jupiter Penninus, while the whole range was called Mons Jovis, a name, under the corrupt form of Mont Joux, which it retained until comparatively recent times. After this the pass became one of the principal highways from Rome to the rich and fertile territories beyond the Alps. A substantial Roman road, well paved, was constructed with infinite pains and skill over the mountain, the remains of which may still be seen near the plain of Jupiter. It was used for centuries; and Roman consul and private soldier alike paused at the simple shrine of Jupiter Penninus, and left their offerings there, in gratitude for the protection afforded them. A large number of Roman coins, bronze medals, and fragments of votive brass tablets has been found on this spot, and are now deposited in the small museum of the convent adjoining the refectory. In the fifth century, the barbarian hordes of Goths under Alaric, of Huns under Attila, and of Vandals under Genseric, swept over the pass to subdue Italy and take possession of Rome. From that time, no event of importance, with the exception of the passage of the Lombards in 547, occurred in connexion with this spot, until Bernhard, who is supposed by some to have given his name to the pass, uncle of Charlemagne, marched a large army

over it in 773, in his famous expedition against
Astolphus, the last Lombard sovereign but one.
Charlemagne himself afterwards recrossed it at the
head of his victorious troops, after conquering
Didier, the last sovereign of Upper Italy. Then
came Bernard de Menthon, in the year 962, and,
abolishing the last remains of Pagan worship,
founded the Hospice which has received his name,
and erected the first Christian altar. After this
period, as Mr. King, in his delightful book, " The
Italian Valleys of the Alps," informs us, the Saracens ravaged the convent, and destroyed its records
by fire, and were in turn attacked and repulsed by
the Normans. Humbert "the white-handed" led
over the pass an army in 1034, to join Conrad in
the conquest of Burgundy; and a part of the army
of Frederic Barbarossa crossed in 1166, under the
command of Berthold de Zähringen. " Pilgrims
bound to Rome frequented it, travelling in large
caravans for mutual protection from the brigands
who infested it after the Saracen invasion; and we
find our own King Canute, himself a pilgrim to the
tomb of St. Peter's, by his representations to the
Pope and the Emperor Adolphus on behalf of his
English pilgrim subjects, obtaining the extirpation
of those lawless bands, and the free and safe use of
the pass." The present building was erected about
the year 1680, its predecessor having been burnt to
the ground. It is impossible to enumerate within
our narrow limits all the remarkable historical
events which are connected with this place, from

the February of the year 59, when Cæcina, the Roman general, marched over it with the cohorts recalled from Britain, through a snow-storm in February, to the spring of the year 1800, when Napoleon crossed it with an army of 80,000 men and 58 field-pieces on his way to the famous battle-field of Marengo. There are few spots in the world that have witnessed so many changes and revolutions, few spots which have been trodden by so many human feet; and I do not envy the man who can gaze upon the narrow path that skirts the lake from the Hospice calm and unmoved, when he thinks of the myriads of his fellow-creatures, from the greatest names in all history down to the lowest and most obscure, who, age after age, have disturbed the stern silence of these rocks, and who have now all alike gone down into undistinguishable dust. Methinks the history of this little footpath is a commentary upon the nothingness of human pride, more impressive than all that poetry has ever sung or philosophy taught!

A little way beyond the Hospice, on a slightly rising ground, is a low building of one storey, built in the rudest manner, and with the roughest materials. It is covered with a grey-slated roof; and in the wall of the gable which fronts you there is a narrow iron grating, through which the light shines into the interior. You look in, and never till your dying day will you forget the ghastly spectacle that then meets your eye. It haunted me like a dreadful nightmare long afterwards. This is the famous Morgue, or dead-house, of which all the world has heard, and which every one visiting the convent, whose nerves are sufficiently strong, makes a special point of seeing. I could almost have wished, however, that my curiosity had been less keen; for it is not pleasant to hang up in the gallery of one's memory a picture like that. And yet it does one good to see it. It softens the heart with pity; it conveys, in a more solemn form than we are accustomed to read it, the lesson of mortality; and it gives us a better idea than we could otherwise have formed of the dangers and sufferings which have often to be encountered in the winter passage of these mountains, and the noble work which the monks of St. Bernard perform. It was indeed a Golgotha, forcibly reminding me of Ezekiel's vision of the valley of dry bones. Skulls, ribs, vertebræ, and other fragments of humanity, with the flesh long ago wasted away from them, blanched by sun and frost, lay here and there in heaps on the floor. As my eye got accustomed to the obscurity of the place, I noticed beyond this mass of miscellaneous bones, separated by a low wall which did not obstruct the view, an extraordinary group of figures. These were the bodies found entire of those who had perished in the winters' snow-storms. Some were lying prostrate, others were leaning against the rough wall, the dim, uncertain light imparting to their faces a strange and awful expression of life. Three figures especially attracted and riveted my attention. In the right-hand corner there was a tall spectre fixed in an upright attitude, with its skeleton arms outstretched, as if supplicating for the aid that never came, and its eyeless sockets glaring as if with a fearful expression. For years it had stood thus without any perceptible change. In another corner there was a figure kneeling upon the floor, muffled in a thick dark cloak, with a blue worsted cuff on the left wrist. No statue of the Laocöon ever told its tale of suffering more eloquently than did that shrivelled corpse. He was an honest and industrious workman, a native of Martigny. He set out early one December morning from that town, intending to go over into Italy in search of employment. He got safely and comfortably as far as the Cantine de Proz, where he halted all night. Next morning he set out through the defile leading up to the Hospice. The weather was at first favourable, but he had not proceeded far when dark clouds speedily covered the sky from end to end, and the fearful *guxen*, which always rages most violently in the Alpine passes, broke out in all its fury. He had doubtless fought against it with all his energy, but in vain. He was found, not three hundred yards from the convent door, buried among the deep snow, frozen in the attitude in which he still appeared, with his knees bent, and his head thrown back in hopeless exhaustion and despair. But the saddest of all the sad sights of the Morgue is the corpse of a woman lying huddled up at the foot of the last-mentioned figure, dressed in dark rags. In her arms she holds a bundle, which you are told is a baby; and her withered face bends over it with a fond expression which death and decay have not been able to obliterate. The light shines full on her quiet features, which are no more ruffled by earthly pain. You cannot fail to see that she had made every effort to preserve the life of the baby to the last moment, for most of her own scanty clothing is drawn up and wrapt round its tiny form, leaving her own limbs exposed to the blast. Oh, sacred mystery of mother's love, stronger than pain, more enduring than death! But alas! in vain was its self-sacrificing tenderness here. The weary feet could no longer bear the precious burden over the wild, and sinking in the fatal sleep, the snow drifted over them, fold by fold, silent and swift, and the place that knew them once knew them no more for ever: the wind passed over it, and it was gone. They found the hapless pair in the following spring, when the snows had melted away; and they bore them tenderly and sadly to this last resting-place. No one came to claim them. Where the poor woman came from, what was her name, no one ever knew; and in this heart-touching pathos of mystery and death she awaits the coming of that other and brighter spring that shall melt even the chill of the tomb.

It is indeed a strange place, that Morgue! All its ghastly tenants perished in the same dreadful way,—the victims of the storm-fiend. Side by side they repose, so cold, so lonely, so forsaken; with no earth to cover them; no token of love from those who were nearest and dearest; no flower to bloom over their dust; not even one green blade of grass to draw down the sunshine and the dew of heaven to their dark charnel-house. Traveller after traveller from the ends of the earth comes and looks in with shuddering dread through the grating on the pitiable sight, and then goes away, perhaps a sadder and a wiser man. For my own part I could not resist the tender impulse, which moved me to gather a small nosegay of gentians and blue-bells, and throw it in, as an offering of pity, to the poor deserted and forgotten dead. It is impossible to dig a grave in this spot, for the hard rock comes everywhere to the surface, and but the thinnest sprinkling of mould rests upon it, hardly sufficient to maintain the scanty vegetation. This sterile region refuses even a grave to those who die there! So cold and dry is the air, that the corpses in the Morgue do not decompose in the same way that they do at lower elevations. They wither and collapse into mummies, embalmed by the air, like the dried bodies preserved in the catacombs of Palermo,—and for years they undergo no change, —at last falling to pieces, and strewing the ground with their fragments. Within the last twelve years no less than sixteen persons have perished in the snow. Some five or six years ago, two of the monks went out with a couple of servants to search for a man who was supposed to have lost himself in the mountains. They were scarcely fifty paces away from the Hospice, when an immense avalanche fell from the side of Mont Chenaletta, and buried the whole party under eighteen feet of snow. The dreadful catastrophe was seen from the convent door, but the monks were utterly powerless to render help. When rescued, the party were all dead. The number of accidents on the St. Bernard pass has greatly diminished of late years; and now the services of the monks in winter are principally required to nurse poor travellers exhausted by the difficulties of the ascent, or who have been frost-bitten. Returning from my morning walk, I saw the famous *marons*, or St. Bernard dogs, playing about the convent door. There were five of them, massively built creatures, of a brown colour,—very like Newfoundland dogs, only larger and more powerful. The stock is supposed to have come originally from the Pyrenees. The services they have rendered in rescuing travellers are incalculable. A whole book might easily be filled with interesting adventures of which they were the heroes. In the Museum at Berne I saw the stuffed body of the well-known dog "Barry," which is said to have saved the lives of no less than forty persons. The huge creatures were fond of being caressed; and one of them ran after my companion, as he was going up the hill-side by a wrong path, and pulled him back by the coat-tail!

After a substantial breakfast, we paid a visit to the chapel to deposit our alms in the alms-box, for though the monks make no charge for their hospitality, or even give the least hint of a donation, there is a box placed in the chapel for the benefit of the poor, and to this fund every traveller should contribute, at the very least, what the same accommodation would have cost him at an hotel. It is to be feared, however, that the great majority contribute nothing at all. Not one of the company who supped and breakfasted with us approached the chapel, having skulked away as soon as they could decently take leave; and yet they were bedizened with gold chains and jewellery of a costly description. There was one Scotchman present who carried out his sound Protestant principles at the expense of the poor monks. He was a very thin, wiry man, but he ate an enormous supper and breakfast. He drank a bottle of wine at each meal, and helped himself most largely to everything on the table. He took what would have sufficed for four ordinary men, and, to our intense disgust, he rubbed down his stomach complacently in the morning ere departing, and said, in the hearing of all, that "he had made up his mind to put nothing in the alms-box, lest he should countenance Popery!" The expenses of the establishment are very heavy, while the funds to meet them have been decreasing. Formerly the convent was the richest in Europe, possessing no less than eighty benefices. But Charles Emmanuel III. of Sardinia, falling into a dispute with the Cantons of Switzerland about the nomination of a provost, sequestrated the possessions of the monks, leaving them only a small estate in the Valais and in the Canton de Vaud. The French and Italian governments give an annual subsidy of a thousand pounds, while another thousand is raised by the gifts of travellers, and by collections made in Switzerland,—Protestants contributing as freely as Roman Catholics. Notwithstanding their comparative poverty, however, the monks are still as lavish and hospitable as ever, up to their utmost means. As it was the feast of the Assumption of the Virgin, crowds of beggars and tramps from the neighbouring valleys,—masses of human degradation and deformity of the most disgusting character,—were congregated about the kitchen door, clamorous for alms, while the monks were busy serving them with bread, cold meat, and wine. What they could not eat they carried away in baskets which they had brought for the purpose. Entering the chapel with our little offering, we were greatly struck with its magnificence, as contrasted with the excessive plainness of the outside, and the sterility of the spot. It is considered a very sacred place, for it contains the relics of no less than three famous saints, viz. St. Bernard, St. Hyrenæus, and St. Maurice, of the celebrated Theban legion of Christians. Five massive gilt altars stood in various parts of the chapel, while the walls were adorned with frescoes and several fine paintings and statues. The marble tomb of Desaix, representing him in relief, wounded and sinking from his horse into the arms of his aide, Le

Brun, was a conspicuous object. "I will give you the Alps for your monument," said Napoleon, with tears in his eyes, to his dying friend: "you shall rest on their loftiest inhabited point." The body of the general was carefully embalmed at Milan, and afterwards conveyed to the chapel, where it now reposes. A crowd of peasants, men and women, were kneeling, during our visit, in the body of the church, performing their devotions; while three or four monks, dressed in splendid habiliments of crimson and gold, were chanting "the solemn melodies of a Gregorian mass," accompanied by the rich tones of a magnificent organ; and clouds of fragrant incense rose slowly to the roof.

Anxious to see the geographical bearings of the convent, we climbed up, with immense expenditure of breath and perspiration, a lofty precipitous peak close at hand. We had a most glorious view from the top, for the atmosphere was perfectly clear, and the remotest distances plainly visible. In front was "le Mont Blanc," as the inhabitants proudly call it, and at this distance of fifteen miles in a straight line it looked infinitely higher and grander than when seen from the nearer and more commonly visited points of view at Chamouni. Far up, miles seemingly, in the deep blue sky, rose the dazzling whiteness of its summit, completely dwarfing all the other peaks around it. On our left was the enormously vast group of Monte Rosa, its everlasting snows tinged with the most delicate crimson hues of the rising sun; while between them the stupendous obelisk of the Matterhorn, by far the sharpest and sublimest of the peaks of Europe, stormed the sky, with a long grey cloud flying at its summit like a flag of defiance. Around these three giant mountains crowded a bewildering host of other summits, most of them above 13,000 feet high, with enormous glaciers streaming down their sides, and forming the sources of nearly all the great rivers of the Continent. My eye and soul turned away from this awful white realm of death, with relief, to the brown and green mountains of Italy, which just peered timidly, as it were, above that fearful horizon in the far south, with an indescribably soft, warm sky brooding over them, as if in sympathy. That little strip of mellow sky and naturally-coloured earth was the only bond in all the wide view that united me to the cosy, lowly world of my fellow-creatures. On this hill, composed of very friable schistose rock, I gathered a considerable number of very interesting plants peculiar to the Alps. The *Arnica montana* displayed its large yellow composite flowers in the shady recesses of the rocks; and, as if to illustrate the proverb that the antidote is ever beside the evil, I found its juicy stems very serviceable in healing a bruise on the leg which I got from a falling stone when gathering specimens. Another composite plant, the *Chrysanthemum alpinum*, whitened in thousands the slopes of *débris*. It has been observed, with *Phyteuma pauciflora*, beside the Lys glacier on Monte Rosa, at 11,352 feet. Nothing could exceed the beauty and luxuriance of the patches of *Linaria alpina*, covered with a profusion of orange and purple labiate blossoms, which spread everywhere over the loose soil. No less striking were the sheets of forget-me-not-like flowers, blue as the sky itself, produced by the *Eritrichium nanum*, growing in the moist sunny fissures. At the base of the hill on the Italian side, where there was a slight tinge of grassy verdure, the yellow star of Bethlehem (*Ornithogalum fistulosum*) and the *Nigritella angustifolia* struggled into existence. The former rises an inch or two above the soil, and produces two or three brilliantly-yellow flowers on each stem; while the compact showy heads of deep blackish crimson flowers of the latter, springing from very short and very narrow leaves, diffuse a fine vanilla-like fragrance. At lower elevations they grow in great profusion, and form the finest ornaments of the Alpine pastures. Among the saxifrages which I observed growing more or less plentifully were the *S. androsacea* (of which I could get no specimen perfect, for the marmot is so fond of it that it nibbles its stems, leaves, and flowers all round), the *S. bryoides, Aizoon, biflora, cæsia*, and *muscoides*. A short distance below the summit there were several large snow-wreaths. Their perpetual drip nourished a glowing little colony of the unrivalled *Gentiana bavarica*, and the compact sheets of the *Androsace glacialis*, sprinkled over with bright pink solitary flowers. In one place there was a curious natural conservatory. The under surface of the snow having been melted by the warmth of the soil—which in Alpine regions is always markedly higher than that of the air—was not in contact with it. A snowy vault was thus formed, glazed on the top with thin plates of transparent ice; and here grew a most lovely cushion of the *Aretia Helvetica*, covered with hundreds of its delicate rosy flowers, like a miniature hydrangea blossom. The dark colour of the soil favoured the absorption of heat; and, prisoned in its crystal cave, this little fairy grew and blossomed securely from the very heart of winter, the unfavourable circumstances around all seeming so many ministers of good, increasing its strength and enhancing its loveliness. Owing to the high temperature of the soil in the Alps, plants are enabled to thrive at great altitudes; and even animal life is not unfrequent at a height of 10,000 feet. I observed at the foot of the snow-wreaths on this hill numerous burrows of a kind of mouse called *Arvicola nivalis*, which is also found on the top of the Faulhorn, Rothhorn, and on the Grands Mulets. Under the stones on the surface of the snow were lively masses of the small, black glacier flea (*Desoria glacialis*); while several specimens of that magnificent butterfly, the *Parnassius Apollo*, distinguished by its white almost transparent wings, marked with scarlet and black-ringed *ocelli*, sailed past with astonishing swiftness in the bright sunshine. These were very satisfactory representatives of the rich animal world we had left behind in the valleys. After a reasonable time spent in the enjoyment of all these treasures, we turned to depart. Hurriedly descending, with many a picturesque tumble and glissade, which did not improve the continuity of our clothing, we reached the foot of the hill in safety. Shortly afterwards we bade adieu to our hospitable entertainers with mingled feelings of gratification and regret: gratification, because we had seen so much that was new and interesting to us, and had been so kindly treated, though strangers in a strange land; and regret, because the palmiest days of the Hospice are over, for the railway tunnel through Mont Cenis, which will soon be completed, will whirl away travellers direct into Italy, and few will care to turn aside, on a long and somewhat difficult journey, to visit the spot.

THE END.

BIBLE TEACHINGS IN NATURE.

THIRD EDITION.

BY THE SAME AUTHOR.

Price 6s.

OPINIONS OF THE PRESS.

"Ably and eloquently written. It is a thoughtful book, and one that is prolific of thought."—*Pall Mall Gazette.*

"It is an attempt to show, by a series of well-selected examples, how all nature, viewed in the light shed upon it by the last-lit lamps of science, sustains and upholds the general teaching, and in many cases even the special expressions, of Holy Writ. The miracles of Scripture are in this view but samples and condensations of the perpetual miracles of nature. This position is illustrated by a copious wealth of instances and an easy flow of poetic language. Mr. Macmillan writes extremely well, and has produced a book which may be fitly described as one of the happiest efforts for enlisting physical science in the direct service of religion. Under his treatment she becomes the willing handmaid of an instructed and contemplative devotion."—*Guardian.*

"We part from Mr. Macmillan with exceeding gratitude. He has made the world more beautiful to us, and unsealed our ears to voices of praise and messages of love that might otherwise have been unheard. We commend the volume, not only as a valuable appendix to works on natural theology, but as a series of prose idylls of unusual merit."—*British Quarterly Review.*

"Mr. Macmillan's eloquent volume is a good one in the broadest sense, and everybody must take pleasure in such talk of the world about us."—*Examiner.*

"The whole work is so pleasant and suggestive that we are sure it must do good to its readers."—*Spectator.*

"No mere extracts will do justice to such a book as this, marked as it is throughout by a singular variety, fulness, and aptitude of illustration, and showing the author's intimate acquaintance, not simply with the teachings of science, but with spiritual truth. . . . For copiousness, vivacity, and beauty of illustration, and for freshness and vigour of higher teaching, it would be hard to find its equal."—*Church and State Review.*

"The writing of the book is most striking, and in many places highly eloquent."—*Literary Churchman.*

"As a contribution to natural theology, though not so expressly intended, we regard the work as very valuable. The author has fine powers of analysis, generalization, and description. He has also a fine eye for analogies, and makes his reader see them as clearly as he does himself. . . . We have said enough to commend this volume to thoughtful readers."—*Record.*

"The descriptions are simple, beautiful, and real. A great deal of information in natural science and many interesting facts of natural history are embodied in them, and the moral and spiritual truths illustrated are for the most part suggested, they are never obtruded upon the reader. It is this which makes the volume much more than pleasant reading. . . . With great heartiness we commend the book to our readers, as containing much information and poetic suggestion, imparted in a simple, clear, and attractive style."—*Nonconformist.*

"A work unrivalled for its unique and harmonious combination of science, poetry, and religion."—"*Piccadilly Papers by a Peripatetic:*" *London Society.*

"Throughout the entire book there is a keen appreciation of what is true in science, and the author possesses a mind well trained to patient research, and capable of investigating the most subtle analogies between the truths of nature and revelation. It is a well-spring of fresh and beautiful thought, and every page is suggestive of some topic for devout meditation. It is not a book for the shallow sciolist either in science or religion, but to those who feel that truth, though many-sided, is yet one and eternal, we promise many pleasant hours in the perusal of these pages."—*Independent.*

"An earnest, thoughtful, graceful book, uniting the poetical with the divine, and lighting with holy, far-seeing fancy the beautiful and the useful as types, evidences, and promises of the wisdom and mercy of God. It is a book that evidences a soul as well as a mind, a deep acknowledgment of light, love, and wisdom beyond all the marvels of science or the quibbles of philosophy. It is eloquent, too, as well as earnest, and none can rise from its perusal untouched by the high and tender truths it inculcates."—*Atlas.*

OPINIONS OF THE PRESS.

"The author of 'First Forms of Vegetation' has fortunately been encouraged to come forward again. He comes in the capacity for which he is eminently fitted, of a teacher whose chief design it is to lead his readers up from nature to nature's God. . . . An exquisite book."—*Illustrated London News.*

"The volume is full of freshness and originality, and everywhere gives evidence of a mind that has loved to study both the book of Nature and the book of Revelation, and to bring each into the light of the other."—*Glasgow Herald.*

"No more delightful volume has for long been added to our shelves than that which is now before us. We know of nothing exactly like it in the language. Hervey was spiritual, but weak and tawdry; Paley, philosophical, but cold: the author of the 'Bible Teachings in Nature' is as positive and ardent in his piety as Hervey, and as philosophical as Paley, without his peculiar hardness. . . . No one who takes up this volume will find it so easy to lay it down."—*Daily Review.*

"Two features of this work give it unusual claims upon the reader's attention. The first is the warmth and vivacity of its style, and the vivid and graphic character of its descriptions. Its scientific fidelity is a second marked characteristic of the work. Its author has cultivated original inquiries in some of the most interesting departments of science, and the book bears the invaluable stamp of a mind which has matured its topics, and is at home in their handling and presentation. The volume is not only filled with agreeable reading, but is rich in instructive scientific exposition."—Dr. YOUMANS, in *New York Independent.*

"It is a long time since the writer has read with such intense satisfaction so interesting and profitable a book. It compels its readers to think. We earnestly recommend it to our readers."—*South American Missionary Magazine.*

"It is a collection of Biblical papers of remarkable freshness and interest. Mr. Macmillan possesses a rare gift both of insight and exposition; both greatly aided by his extensive acquaintance with natural science.. The more obvious Bible symbols from nature acquire at his hands a fulness, point, and richness like that which the microscope gives to familiar objects; while many that are obscure to the ordinary reader start from his touch flashing with light and beauty."—Professor BLAIKIE, in *Sunday Magazine.*

www.ingramcontent.com/pod-product-compliance
Lightning Source LLC
Chambersburg PA
CBHW031905220426
43663CB00006B/782